Waste Management

Waste Management

Edited by
Mateo Roberts

Larsen & Keller
www.larsen-keller.com

Waste Management
Edited by Mateo Roberts
ISBN: 978-1-63549-287-3 (Hardback)

© 2017 Larsen & Keller

Larsen & Keller

Published by Larsen and Keller Education,
5 Penn Plaza,
19th Floor,
New York, NY 10001, USA

Cataloging-in-Publication Data

Waste management / edited by Mateo Roberts.
    p. cm.
Includes bibliographical references and index.
ISBN 978-1-63549-287-3
1. Refuse and refuse disposal. 2. Salvage (Waste, etc.). 3. Recycling (Waste, etc.). 4. Waste minimization.
I. Roberts, Mateo.
TD791 .W37 2017
628.44--dc23

The publisher's policy is to use permanent paper from mills that operate a sustainable forestry policy. Furthermore, the publisher ensures that the text paper and cover boards used have met acceptable environmental accreditation standards.

Printed and bound in the United States of America.

For more information regarding Larsen and Keller Education and its products, please visit the publisher's website www.larsen-keller.com

Table of Contents

Preface

As the population of the world is rising, so is the amount of waste being produced. The mismanagement of waste has led to serious ecological damage as well as social and environmental harm. Waste management refers to the field of using various actions, activities, technologies and methods to manage waste. It also includes transport, collection, disposal and treatment of waste. This book attempts to understand the multiple branches that fall under the discipline of waste management and how such concepts have practical applications. Most of the topics introduced in this book cover new techniques and the methods of this subject. Different approaches, evaluations and advanced studies have been included in it. For all those who are interested in waste management, this textbook can prove to be an essential guide.

To facilitate a deeper understanding of the contents of this book a short introduction of every chapter is written below:

Chapter 1- The management of waste, from its beginning to its final removal is known as waste management. It also deals with recycling and collection and transportation of waste. Waste management helps in reducing the effects caused by waste. This chapter on waste management offers an insightful focus, keeping in mind the complex subject matter.

Chapter 2- Unwanted substances and materials are termed as waste. Any substance which is cast-off after being used is regarded as waste. Some of the types of waste are chemical waste, municipal solid waste, biomedical waste, commercial waste etc. There is another type of waste known as biodegradable waste, it includes substances that can be decomposed and are usually items such as food waste, manure and swage sludge.

Chapter 3- Waste needs to be handled in order to keep our environment and our surroundings clean. There are a number of practices that handle waste; some of these are kerbside collection, garbage truck, sand cleaning machine and waste collection. The service provided to urban cities for removing waste is termed as kerbside collection. The chapter strategically encompasses and incorporates the major components and key concepts of waste management, providing a complete understanding.

Chapter 4- This chapter focuses on different practices of disposing waste. Some of these practices are landfill, incineration and compost. Waste materials have certain organic substances. The burning of these organic substances is known as incineration. Landfills on the other hand are locations, which are used for the removal of waste materials by burial.

Chapter 5- A number of strategies are used to manage waste. Some of the strategies are co-processing, curb mining and land farming. Waste that can be used as raw material or as a source of energy is referred to as co-processing. These strategies teach the recycling and usage of waste in whichever manner possible, to save resources and energy.

Chapter 6- Water that is gravely affected by human activities is known as wastewater. A common example of wastewater is sewage. A number of techniques have been used to treat sewage. Some of these are aerobic treatment system, trickling filter, sewage treatment and industrial wastewater treatment. This chapter is an overview of the subject matter incorporating all the major aspects of waste management.

Chapter 7- Certain waste material can be reused and the process by which it becomes reusable is known as recycling. Recycling can be of certain types such as plastic recycling, computer recycling, ferrous metal recycling and precycling. The following chapter explains to the reader the importance of recycling.

Chapter 8- The withdrawal of disposed materials for recycling to obtain full benefits is termed as resource recovery and the organizations that promote eco-innovation is industrial symbiosis. Other measures of waste management are eco-industrial park, pyrolysis, waste hierarchy and resource efficiency. The text strategically encompasses and incorporates the major components and key concepts of waste management.

Chapter 9- Every county has specified policies for the management of waste. Some of these policies are solid waste policy in the United States of America, waste management in Turkey and the regional custom of polluter pays principle. The topics discussed in the chapter are of great importance to broaden the existing knowledge on waste management.

Chapter 10- Countries with the purpose of recycling and treatment trade in waste. This trade is regarded as global waste trade. Countries trade in waste for treatment and recycling. This trade is regarded global waste trade. The following chapter concentrates on all the movements related to global waste management.

I owe the completion of this book to the never-ending support of my family, who supported me throughout the project.

Editor

Introduction to Waste Management

The management of waste, from its beginning to its final removal is known as waste management. It also deals with recycling and collection and transportation of waste. Waste management helps in reducing the effects caused by waste. This chapter on waste management offers an insightful focus, keeping in mind the complex subject matter.

Waste management is all the activities and actions required to manage waste from its inception to its final disposal. This includes amongst other things, collection, transport, treatment and disposal of waste together with monitoring and regulation. It also encompasses the legal and regulatory framework that relates to waste management encompassing guidance on recycling etc.

Waste management in Kathmandu, Nepal

The term normally relates to all kinds of waste, whether generated during the extraction of raw materials, the processing of raw materials into intermediate and final products, the consumption of final products, or other human activities, including municipal (residential, institutional, commercial), agricultural, and social (health care, household hazardous wastes, sewage sludge). Waste management is intended to reduce adverse effects of waste on health, the environment or aesthetics.

Waste management practices are not uniform among countries (developed and developing nations); regions (urban and rural area), and sectors (residential and industrial).

Central Principles of Waste Management

There are a number of concepts about waste management which vary in their usage between countries or regions. Some of the most general, widely used concepts include:

most favoured option

prevention

minimisation

reuse

recycling

energy recovery

least favoured option

disposal

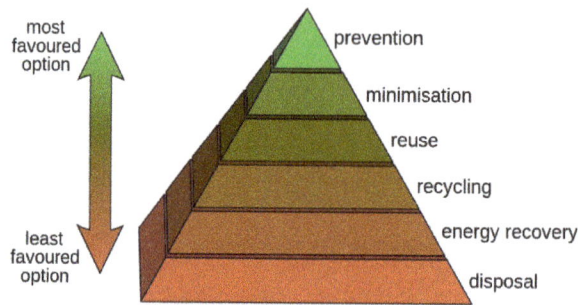

Diagram of the waste hierarchy

Waste Hierarchy

The waste hierarchy refers to the "3 Rs" reduce, reuse and recycle, which classify waste management strategies according to their desirability in terms of waste minimisation. The waste hierarchy remains the cornerstone of most waste minimisation strategies. The aim of the waste hierarchy is to extract the maximum practical benefits from products and to generate the minimum amount of waste; see: resource recovery.The waste hierarchy is represented as a pyramid because the basic premise is for policy to take action first and prevent the generation of waste. The next step or preferred action is to reduce the generation of waste i.e. by re-use. The next is recycling which would include composting. Following this step is material recovery and waste-to-energy. Energy can be recovered from processes i.e. landfill and combustion, at this level of the hierarchy. The final action is disposal, in landfills or through incineration without energy recovery. This last step is the final resort for waste which has not been prevented, diverted or recovered. The waste hierarchy represents the progression of a product or material through the sequential stages of the pyramid of waste management. The hierarchy represents the latter parts of the life-cycle for each product.

Life-cycle of a Product

The life-cycle begins with design, then proceeds through manufacture, distribution, use and then follows through the waste hierarchy's stages of reuse, recovery, recycling and disposal. Each of the above stages of the life-cycle offers opportunities for policy intervention, to rethink the need for the product, to redesign to minimize waste potential, to extend its use. The key behind the life-cycle of a product is to optimize the use of the world's limited resources by avoiding the unnecessary generation of waste.

Resource Efficiency

Resource efficiency reflects the understanding that current, global, economic growth and development can not be sustained with the current production and consumption patterns. Globally, we are extracting more resources to produce goods than the planet can replenish. Resource efficiency is the reduction of the environmental impact from the production and consumption of these goods, from final raw material extraction to last use and disposal. This process of resource efficiency can address sustainability.

Polluter Pays Principle

The Polluter pays principle is a principle where the polluting party pays for the impact caused to the environment. With respect to waste management, this generally refers to the requirement for a waste generator to pay for appropriate disposal of the unrecoverable material.

History

Throughout most of history, the amount of waste generated by humans was insignificant due to low population density and low societal levels of the exploitation of natural resources. Common waste produced during pre-modern times was mainly ashes and human biodegradable waste, and these were released back into the ground locally, with minimum environmental impact. Tools made out of wood or metal were generally reused or passed down through the generations.

However, some civilizations do seem to have been more profligate in their waste output than others. In particular, the Maya of Central America had a fixed monthly ritual, in which the people of the village would gather together and burn their rubbish in large dumps.

Modern Era

Sir Edwin Chadwick's 1842 report *The Sanitary Condition of the Labouring Population* was influential in securing the passage of the first legislation aimed at waste clearance and disposal.

Following the onset of industrialisation and the sustained urban growth of large population centres in England, the buildup of waste in the cities caused a rapid deterioration in levels of sanitation and the general quality of urban life. The streets became choked with filth due to the lack of waste clearance regulations. Calls for the establishment of a municipal authority with waste removal powers occurred as early as 1751, when Corbyn Morris in London proposed that "...as the preservation of the health of the people is of great importance, it is proposed that the cleaning of this city, should be put under one uniform public management, and all the filth be...conveyed by the Thames to proper distance in the country".

However, it was not until the mid-19th century, spurred by increasingly devastating cholera outbreaks and the emergence of a public health debate that the first legislation on the issue emerged. Highly influential in this new focus was the report *The Sanitary Condition of the Labouring Population* in 1842 of the social reformer, Edwin Chadwick, in which he argued for the importance of adequate waste removal and management facilities to improve the health and wellbeing of the city's population.

In the UK, the Nuisance Removal and Disease Prevention Act of 1846 began what was to be a steadily evolving process of the provision of regulated waste management in London. The Metropolitan Board of Works was the first city-wide authority that centralized sanitation regulation for the rapidly expanding city and the Public Health Act 1875 made it compulsory for every household to deposit their weekly waste in "moveable receptacles: for disposal—the first concept for a dust-bin.

Manlove, Alliott & Co. Ltd. 1894 destructor furnace. The use of incinerators for waste disposal became popular in the late 19th century.

The dramatic increase in waste for disposal led to the creation of the first incineration plants, or, as they were then called, "destructors". In 1874, the first incinerator was built in Nottingham by Manlove, Alliott & Co. Ltd. to the design of Albert Fryer. However, these were met with opposition on account of the large amounts of ash they produced and which wafted over the neighbouring areas.

Similar municipal systems of waste disposal sprung up at the turn of the 20th century in other large cities of Europe and North America. In 1895, New York City became the first U.S. city with public-sector garbage management.

Early garbage removal trucks were simply open bodied dump trucks pulled by a team of horses. They became motorized in the early part of the 20th century and the first close body trucks to eliminate odours with a dumping lever mechanism were introduced in the 1920s in Britain. These were soon equipped with 'hopper mechanisms' where the scooper was loaded at floor level and then hoisted mechanically to deposit the waste in the truck. The Garwood Load Packer was the first truck in 1938, to incorporate a hydraulic compactor.

Waste Handling and Transport

Waste collection methods vary widely among different countries and regions. Domestic waste collection services are often provided by local government authorities, or by private companies for industrial and commercial waste. Some areas, especially those in less developed countries, do not have formal waste-collection systems.

Molded plastic, wheeled waste bin in Berkshire, England

Waste Handling Practices

Curbside collection is the most common method of disposal in most European countries, Canada, New Zealand and many other parts of the developed world in which waste is collected at regular intervals by specialised trucks. This is often associated with curb-side waste segregation. In rural areas waste may need to be taken to a transfer station. Waste collected is then transported to an appropriate disposal facility. In some areas, vacuum collection is used in which waste is transported from the home or commercial premises by vacuum along small bore tubes. Systems are in use in Europe and North America.

Pyrolysis is used to dispose of some wastes including tires, a process that can produce recovered fuels, steel and heat. In some cases tires can provide the feedstock for cement manufacture. Such systems are used in USA, California, Australia, Greece, Mexico, the United Kingdom and in Israel. The RESEM pyrolysis plant that has been operational at Texas USA since December 2011, and processes up to 60 tons per day. In some jurisdictions unsegregated waste is collected at the curbside or from waste transfer stations and then sorted into recyclables and unusable waste. Such systems are capable of sorting large volumes of solid waste, salvaging recyclables, and turning the rest into bio-gas and soil conditioner. In San Francisco, the local government established its Mandatory Recycling and Composting Ordinance in support of its goal of zero waste by 2020, requiring everyone in the city to keep recyclables and compostables out of the landfill. The three streams are collected with the curbside "Fantastic 3" bin system - blue for recyclables, green for compostables, and black for landfill-bound materials - provided to residents and businesses and serviced by San Francisco's sole refuse hauler, Recology. The City's "Pay-As-You-Throw" system charges customers by the volume of landfill-bound materials, which provides a financial incentive to separate recyclables and compostables from other discards. The City's Department of the Environment's Zero Waste Program has led the City to achieve 80% diversion, the highest diversion rate in North America. Other businesses such as Waste Industries use a variety of colors to distinguish between trash and recycling cans.

Financial Models

In most developed countries, domestic waste disposal is funded from a national or local tax which may be related to income, or notional house value. Commercial and industrial waste disposal is typically charged for as a commercial service, often as an integrated charge which includes disposal costs. This practice may encourage disposal contractors to opt for the cheapest disposal option such as landfill rather than the environmentally best solution such as re-use and recycling. In some areas such as Taipei, the city government charges its households and industries for the volume of rubbish they produce. Waste will only be collected by the city council if waste is disposed in government issued rubbish bags. This policy has successfully reduced the amount of waste the city produces and increased the recycling rate.

Disposal Solutions

Landfill

A landfill compaction vehicle in action.

Spittelau incineration plant in Vienna

Incineration

Incineration is a disposal method in which solid organic wastes are subjected to combustion so as to convert them into residue and gaseous products. This method is useful for disposal of residue of both solid waste management and solid residue from waste water management. This process reduces the volumes of solid waste to 20 to 30 percent of the original volume. Incineration and other high temperature waste treatment systems are sometimes described as "thermal treatment". Incinerators convert waste materials into heat, gas, steam, and ash.

Incineration is carried out both on a small scale by individuals and on a large scale by industry. It is used to dispose of solid, liquid and gaseous waste. It is recognized as a practical method of disposing of certain hazardous waste materials (such as biological medical waste). Incineration is a controversial method of waste disposal, due to issues such as emission of gaseous pollutants.

Incineration is common in countries such as Japan where land is more scarce, as these facilities generally do not require as much area as landfills. Waste-to-energy (WtE) or energy-from-waste (EfW) are broad terms for facilities that burn waste in a furnace or boiler to generate heat, steam or electricity. Combustion in an incinerator is not always perfect and there have been concerns about pollutants in gaseous emissions from incinerator stacks. Particular concern has focused on some very persistent organic compounds such as dioxins, furans, and PAHs, which may be created and which may have serious environmental consequences.

Recycling

Recycling is a resource recovery practice that refers to the collection and reuse of waste materials such as empty beverage containers. The materials from which the items are made can be reprocessed into new products. Material for recycling may be collected separately from general waste using dedicated bins and collection vehicles, a procedure called kerbside collection. In some communities, the owner of the waste is required to separate the materials into different bins (e.g. for paper, plastics, metals) prior to its collection. In other communities, all recyclable materials are placed in a single bin for collection, and the sorting is handled later at a central facility. The latter method is known as "single-stream recycling."

Steel crushed and baled for recycling

The most common consumer products recycled include aluminium such as beverages cans, copper such as wire, steel from food and aerosol cans, old steel furnishings or equipment, rubber tyres, polyethylene and PET bottles, glass bottles and jars, paperboard cartons, newspapers, magazines and light paper, and corrugated fiberboard boxes.

PVC, LDPE, PP, and PS are also recyclable. These items are usually composed of a single type of material, making them relatively easy to recycle into new products. The recycling of complex products (such as computers and electronic equipment) is more difficult, due to the additional dismantling and separation required.

Waste not the Waste. Sign in Tamil Nadu, India

The type of material accepted for recycling varies by city and country. Each city and country has different recycling programs in place that can handle the various types of recyclable materials. However, certain variation in acceptance is reflected in the resale value of the material once it is reprocessed.

Re-use

Biological Reprocessing

Recoverable materials that are organic in nature, such as plant material, food scraps, and paper products, can be recovered through composting and digestion processes to decompose the organic matter. The resulting organic material is then recycled as mulch or compost for agricultural or landscaping purposes. In addition, waste gas from the process (such as methane) can be captured and used for generating electricity and heat (CHP/cogeneration) maximising efficiencies. The intention of biological processing in waste management is to control and accelerate the natural process of decomposition of organic matter.

An active compost heap.

Energy Recovery

Energy recovery from waste is the conversion of non-recyclable waste materials into usable heat, electricity, or fuel through a variety of processes, including combustion, gasification, pyrolyzation, anaerobic digestion, and landfill gas recovery. This process is often called waste-to-energy. Energy recovery from waste is part of the non-hazardous waste management hierarchy. Using energy recovery to convert non-recyclable waste materials into electricity and heat, generates a renewable energy source and can reduce carbon emissions by offsetting the need for energy from fossil sources as well as reduce methane generation from landfills. Globally, waste-to-energy accounts for 16% of waste management.

The energy content of waste products can be harnessed directly by using them as a direct combustion fuel, or indirectly by processing them into another type of fuel. Thermal treatment ranges from using waste as a fuel source for cooking or heating and the use of the gas fuel, to fuel for boilers to generate steam and electricity in a turbine.

Pyrolysis and gasification are two related forms of thermal treatment where waste materials are heated to high temperatures with limited oxygen availability. The process usually occurs in a sealed vessel under high pressure. Pyrolysis of solid waste converts the material into solid, liquid and gas products. The liquid and gas can be burnt to produce energy or refined into other chemical products (chemical refinery). The solid residue (char) can be further refined into products such as activated carbon. Gasification and advanced Plasma arc gasification are used to convert organic materials directly into a synthetic gas (syngas) composed of carbon monoxide and hydrogen. The gas is then burnt to produce electricity and steam. An alternative to pyrolysis is high temperature and pressure supercritical water decomposition (hydrothermal monophasic oxidation).

Pyrolysis

Pyrolysis is a process of thermo-chemical decomposition of organic materials by heat in the absence of oxygen which produces various hydrocarbon gases. During pyrolysis, the molecules of object are subjected to very high temperatures leading to very high vibrations. Therefore, every molecule in the object is stretched and shaken to an extent that molecules starts breaking down. The rate of pyrolysis increases with temperature. In industrial applications, temperatures are above 430 °C (800 °F). Fast pyrolysis produces liquid fuel for feedstocks like wood. Slow pyrolysis produces gases and solid charcoal. Pyrolysis hold promise for conversion of waste biomass into useful liquid fuel. Pyrolysis of waste plastics can produce millions of litres of fuel. Solid products of this process contain metals, glass, sand and pyrolysis coke which cannot be converted to gas in the process.

Resource Recovery

Resource recovery is the systematic diversion of waste, which was intended for disposal, for a specific next use. It is the processing of recyclables to extract or recover materials and resources, or convert to energy. These activities are performed at a resource recovery facility. Resource recovery is not only environmentally important, but it is also cost effective. It decreases the amount of waste for disposal, saves space in landfills, and conserves natural resources.

Resource recovery (as opposed to waste management) uses LCA (life cycle analysis) attempts to offer alternatives to waste management. For mixed MSW (Municipal Solid Waste) a number of broad studies have indicated that administration, source separation and collection followed by reuse and recycling of the non-organic fraction and energy and compost/fertilizer production of the organic material via anaerobic digestion to be the favoured path.

As an example of how resource recycling can be beneficial, many of the items thrown away contain precious metals which can be recycled to create a profit, such as the components in circuit boards. Other industries can also benefit from resource recycling with the wood chippings in pallets and other packaging materials being passed onto sectors such as the horticultural profession. In this instance, workers can use the recycled chips to create paths, walkways, or arena surfaces.

Sustainability

The management of waste is a key component in a business' ability to maintaining ISO14001 accreditation. Companies are encouraged to improve their environmental efficiencies each year by eliminating waste through resource recovery practices, which are sustainability-related activities. One way to do this is by shifting away from waste management to resource recovery practices like recycling materials such as glass, food scraps, paper and cardboard, plastic bottles and metal. a lot of conferences will discuss this topic as the international Conference on Green Urabnism which will be held in Italy From 12–14 October 2016.

Avoidance and Reduction Methods

An important method of waste management is the prevention of waste material being created, also known as waste reduction. Methods of avoidance include reuse of second-hand products, repairing broken items instead of buying new, designing products to be refillable or reusable (such as cotton instead of plastic shopping bags), encouraging consumers to avoid using disposable products (such as disposable cutlery), removing any food/liquid remains from cans and packaging, and designing products that use less material to achieve the same purpose (for example, lightweighting of beverage cans).

International Waste Movement

While waste transport within a given country falls under national regulations, trans-boundary movement of waste is often subject to international treaties. A major concern to many countries in the world has been hazardous waste. The Basel Convention, ratified by 172 countries, deprecates movement of hazardous waste from developed to less developed countries. The provisions of the Basel convention have been integrated into the EU waste shipment regulation. Nuclear waste, although considered hazardous, does not fall under the jurisdiction of the Basel Convention.

Benefits

Waste is not something that should be discarded or disposed of with no regard for future use. It can be a valuable resource if addressed correctly, through policy and practice. With rational and consistent waste management practices there is an opportunity to reap a range of benefits. Those benefits include:

1. Economic - Improving economic efficiency through the means of resource use, treatment and disposal and creating markets for recycles can lead to efficient practices in the production and consumption of products and materials resulting in valuable materials being recovered for reuse and the potential for new jobs and new business opportunities.

2. Social - By reducing adverse impacts on health by proper waste management practices, the resulting consequences are more appealing settlements. Better social advantages can lead to new sources of employment and potentially lifting communities out of poverty especially in some of the developing poorer countries and cities.

3. Environmental - Reducing or eliminating adverse impacts on the environmental through reducing, reusing and recycling, and minimizing resource extraction can provide improved air and water quality and help in the reduction of greenhouse gas emissions.

4. Inter-generational Equity - Following effective waste management practices can provide subsequent generations a more robust economy, a fairer and more inclusive society and a cleaner environment.

Challenges in Developing Countries

Waste management in cities with developing economies and economies in transition experience exhausted waste collection services, inadequately managed and uncontrolled dumpsites and the problems are worsening. Problems with governance also complicate the situation. Waste management, in these countries and cities, is an ongoing challenge and many struggle due to weak institutions, chronic under-resourcing and rapid urbanization. All of these challenges along with the lack of understanding of different factors that contribute to the hierarchy of waste management, affect the treatment of waste.

Technologies

Traditionally the waste management industry has been a late adopter of new technologies such as RFID (Radio Frequency Identification) tags, GPS and integrated software packages which enable better quality data to be collected without the use of estimation or manual data entry.

References

- United Nations Environmental Programme (2013). "Guidelines for National Waste Management Strategies Moving from Challenges to Opportunities." (PDF). ISBN 978-92-807-3333-4.

- Gandy, Matthew (1994). Recycling and the Politics of Urban Waste. Earthscan. ISBN 9781853831683.

- Barbalace, Roberta Crowell (2003-08-01). "The History of Waste". EnvironmentalChemistry. com. Retrieved 2013-12-09.

- National Waste & Recycling Association. "History of Solid Waste Management". Washington, DC. Retrieved 2013-12-09.

2

Waste and Its Types

Unwanted substances and materials are termed as waste. Any substance which is cast-off after being used is regarded as waste. Some of the types of waste are chemical waste, municipal solid waste, biomedical waste, commercial waste etc. There is another type of waste known as biodegradable waste, it includes substances that can be decomposed and are usually items such as food waste, manure and swage sludge.

Waste

Waste and wastes are unwanted or unusable materials. Waste is any substance which is discarded after primary use, or it is worthless, defective and of no use.

Solid waste being shredded

The term is often bad don't use it subjective (because what is waste to one need not necessarily be waste to another) and sometimes objectively inaccurate (for example, to send scrap metals to a landfill is to inaccurately classify them as waste, because they are recyclable). Examples include municipal solid waste (household trash/refuse), hazardous waste, wastewater (such as sewage, which contains bodily wastes (feces and urine) and surface runoff), radioactive waste, and others.

Definitions

United Nations Environment Program

According to the Basel Convention of 1989, "'Wastes' are substance or objects, which are disposed of or are intended to be disposed of or are required to be disposed of by the provisions of national law"

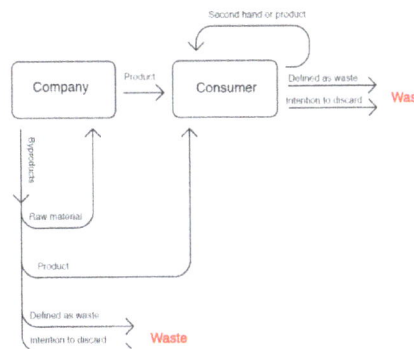

Schematic illustration of the EU Legal definition of waste

United Nations Statistics Division, Glossary of Environment Statistics

"Wastes are materials that are not prime products (that is products produced for the market) for which the initial user has no further use in terms of his/her own purposes of production, transformation or consumption, and of which he/she wants to dispose. Wastes may be generated during the extraction of raw materials, the processing of raw materials into intermediate and final products, the consumption of final products, and other human activities. Residuals recycled or reused at the place of generation are excluded."

European Union

Under the Waste Framework Directive, the European Union defines waste as "an object the holder discards, intends to discard or is required to discard."

Types

There are many waste types defined by modern systems of waste management, notably including:

- Municipal waste includes household waste, commercial waste, and demolition waste

- Hazardous waste includes industrial waste

- Biomedical waste includes clinical waste

- Special hazardous waste includes radioactive waste, explosive waste, and electronic waste (e-waste)

Reporting

There are many issues that surround reporting waste. It is most commonly measured by size or weight, and there is a stark difference between the two. For example, organic waste is much heavier when it is wet, and plastic or glass bottles can have different weights but be the same size. On a global scale it is difficult to report waste because countries have different definitions of waste and what falls into waste categories, as well as different ways of reporting. Based on incomplete reports from its parties, the Basel Convention estimated 338 million tonnes of waste was generated in 2001. For the same year, OECD estimated 4 billion tonnes from its member countries. Despite these inconsistencies, waste reporting is still useful on a small and large scale to determine key causes and locations, and to find ways of preventing, minimizing, recovering, treating, and disposing waste.

Costs

Environmental Costs

Inappropriately managed waste can attract rodents and insects, which can harbour gastrointestinal parasites, yellow fever, worms, the plague and other conditions for humans, and exposure to hazardous wastes, particularly when they are burned, can cause various other diseases including cancers. Toxic waste materials can contaminate surface water, groundwater, soil, and air which causes more problems for humans, other species, and ecosystems. Waste treatment and disposal produces significant green house gas (GHG) emissions, notably methane, which are contributing significantly to global warming.

Social Costs

Waste management is a significant environmental justice issue. Many of the environmental burdens cited above are more often borne by marginalized groups, such as racial minorities, women, and residents of developing nations. NIMBY (not in my back yard) is the opposition of residents to a proposal for a new development because it is close to them. However, the need for expansion and siting of waste treatment and disposal facilities is increasing worldwide. There is now a growing market in the transboundary movement of waste, and although most waste that flows between countries goes between developed nations, a significant amount of waste is moved from developed to developing nations.

Economic Costs

The economic costs of managing waste are high, and are often paid for by municipal governments; money can often be saved with more efficiently designed collection routes, modifying vehicles, and with public education. Environmental policies such as pay as you throw can reduce the cost of management and reduce waste quantities. Waste recovery (that is, recycling, reuse) can curb economic costs because it avoids extracting raw materials and often cuts transportation costs. "Economic assessment of municipal waste management systems – case studies using a combination of life-cycle assessment (LCA) and life-cycle costing (LCC)". The location of waste treatment and disposal facilities often reduces property values due to noise, dust, pollution, unsightliness, and negative stigma. The informal waste sector consists mostly of waste pickers who scavenge for metals, glass, plastic, textiles, and other materials and then trade them for a profit. This sector can significantly alter or reduce waste in a particular system, but other negative economic effects come with the disease, poverty, exploitation, and abuse of its workers.

Resource Recovery

Resource recovery is the retrieval of recyclable waste, which was intended for disposal, for a specific next use. It is the processing of recyclables to extract or recover materials and resources, or convert to energy. This process is carried out at a resource recovery facility. Resource recovery is not only important to the environment, but it can be cost effective by decreasing the amount of waste sent to the disposal stream, reduce the amount of space needed for landfills, and protect limited natural resources.

Energy Recovery

Energy recovery from waste is using non-recyclable waste materials and extracting from it heat, electricity, or energy through a variety of processes, including combustion, gasification, pyrolyzation, and anaerobic digestion. This process is referred to as waste-to-energy.

There are several ways to recover energy from waste. Anaerobic digestion is a naturally occurring process of decomposition where organic matter is reduced to a simpler chemical component in the absence of oxygen. Incineration or direct controlled burning of municipal solid waste to reduce waste and make energy. Secondary recovered fuel is the energy recovery from waste that cannot be reused or recycled from mechanical and biological treatment activities. Pyrolysis involves heating of waste, with the absence of oxygen, to high temperatures to break down any carbon content into a mixture of gaseous and liquid fuels and solid residue. Gasification is the conversion of carbon rich material through high temperature with partial oxidation into a gas stream. Plasma arc heating is the very high heating of municipal solid waste to temperatures ranging from 3,000-10,000 °C, where energy is released by an electrical discharge in an inert atmosphere.

Using waste as fuel can offer important environmental benefits. It can provide a safe and cost-effective option for wastes that would normally have to be dealt with through disposal. It can help reduce carbon dioxide emissions by diverting energy use from fossil fuels, while also generating energy and using waste as fuel can reduce the methane emissions generated in landfills by averting waste from landfills.

There is some debate in the classification of certain biomass feedstock as wastes. Crude Tall Oil (CTO), a co-product of the pulp and papermaking process, is defined as a waste or residue in some European countries when in fact it is produced "on purpose" and has significant value add potential in industrial applications. Several companies use CTO to produce fuel, while the pine chemicals industry maximizes it as a feedstock "producing low-carbon, bio-based chemicals" through cascading use.

Education and Awareness

Education and awareness in the area of waste and waste management is increasingly important from a global perspective of resource management. The Talloires Declaration is a declaration for sustainability concerned about the unprecedented scale and speed of environmental pollution and degradation, and the depletion of natural resources. Local, regional, and global air pollution; accumulation and distribution of toxic wastes; destruction and depletion of forests, soil, and water; depletion of the ozone layer and emission of "green house" gases threaten the survival of humans and thousands of other living species, the integrity of the earth and its biodiversity, the security of nations, and the heritage of future generations. Several universities have implemented the Talloires Declaration by establishing environmental management and waste management programs, e.g. the waste management universityproject. University and vocational education are promoted by various organizations, e.g. WAMITAB and Chartered Institution of Wastes Management.

Municipal Solid Waste

Municipal solid waste (MSW), commonly known as trash or garbage in the United States and as refuse or rubbish in Britain, is a waste type consisting of everyday items that are discarded by the public. "Garbage" can also refer specifically to food waste, as in a garbage disposal; the two are sometimes collected separately.

Composition

The composition of municipal solid waste varies greatly from municipality to municipality (country to country) and changes significantly with time. In municipalities (countries) which have a well developed waste recycling system, the waste stream consists mainly of intractable wastes such as plastic film, and unrecyclable packaging materials. At the start of the 20th century, the majority of domestic waste (53%) in the UK consisted of coal ash from open fires. In developed municipalities (countries) without significant recycling activity it predominantly includes food wastes, market wastes, yard wastes, plastic containers and product packaging materials, and other miscellaneous solid wastes from residential, commercial, institutional, and industrial sources. Most definitions of municipal solid waste do not include industrial wastes, agricultural wastes, medical waste, radioactive waste or sewage sludge. Waste collection is performed by the municipality within a given area. The term *residual waste* relates to waste left from household sources containing materials that have not been separated out or sent for reprocessing. Waste can be classified in several ways but the following list represents a typical classification:

- Biodegradable waste: food and kitchen waste, green waste, paper (most can be recycled although some difficult to compost plant material may be excluded)

- Recyclable materials: paper, cardboard, glass, bottles, jars, tin cans, aluminum cans, aluminum foil, metals, certain plastics, fabrics, clothes, tires, batteries, etc.

- Inert waste: construction and demolition waste, dirt, rocks, debris

- Electrical and electronic waste (WEEE) - electrical appliances, light bulbs, washing machines, TVs, computers, screens, mobile phones, alarm clocks, watches, etc.

- Composite wastes: waste clothing, Tetra Packs, waste plastics such as toys

- Hazardous waste including most paints, chemicals, tires, batteries, light bulbs, electrical appliances, fluorescent lamps, aerosol spray cans, and fertilizers

- Toxic waste including pesticides, herbicides, and fungicides

- Biomedical waste, expired pharmaceutical drugs, etc.

Components of Solid Waste Management

The municipal solid waste industry has four components: recycling, composting, disposal, and waste-to-energy via incineration. There is no single approach that can be applied to the management of all waste streams, therefore the Environmental Protection Agency, federal agency of the United States of America, developed a hierarchy ranking strategy for municipal solid waste. The Waste Management Hierarchy is made up of four levels ordered from most preferred to least preferred methods based on their environmental soundness: Source reduction and reuse; recycling or composting; energy recovery; treatment and disposal.

Bins to collect paper, aluminium, glass, PET bottles and incinerable waste.

Collection

The functional element of collection includes not only the gathering of solid waste and recyclable materials, but also the transport of these materials, after collection, to the location where the collection vehicle is emptied. This location may be a materials processing facility, a transfer station or a landfill disposal site.

Waste Handling and Separation, Storage and Processing at the Source

Waste handling and separation involves activities associated with waste management until the waste is placed in storage containers for collection. Handling also encompasses the movement of loaded containers to the point of collection. Separating different types of waste components is an important step in the handling and storage of solid waste at the source.

Segregation and Processing and Transformation of Solid Wastes

The types of means and facilities that are now used for the recovery of waste materials that have been separated at the source include curbside ('kerbside' in the UK) collection, drop-off and buy-back centers. The separation and processing of wastes that have been separated at the source and the separation of commingled wastes usually occur at a materials recovery facility, transfer stations, combustion facilities and disposal sites.

Transfer and Transport

This element involves two main steps. First, the waste is transferred from a smaller collection vehicle to larger transport equipment. The waste is then transported, usually over long distances, to a processing or disposal site.

Disposal

Today, the disposal of wastes by land filling or land spreading is the ultimate fate of all solid wastes, whether they are residential wastes collected and transported directly to a landfill site, residual materials from materials recovery facilities (MRFs), residue from the combustion of solid waste, compost, or other substances from various solid waste processing facilities. A modern sanitary landfill is not a dump; it is an engineered facility used for disposing of solid wastes on land without creating nuisances or hazards to public health or safety, such as the problems of insects and the contamination of ground water.

Mixed municipal waste, Hiriya, Tel Aviv

Reusing

In the recent years environmental organizations, such as Freegle or Freecycle Network, have been gaining popularity for their online reuse networks. These networks provide a worldwide online registry of unwanted items that would otherwise be thrown away, for individuals and nonprofits to reuse or recycle. Therefore, this free Internet-based service reduces landfill pollution and promotes the gift economy.

Landfills

Landfills are created by land dumping. Land dumping methods vary, most commonly it involves the mass dumping of waste into a designated area, usually a hole or sidehill. After the waste is dumped, it is then compacted by large machines. When the dumping cell is full, it is then "sealed" with a plastic sheet and covered in several feet of dirt. This is the primary method of dumping in the United States because of the low cost and abundance of unused land in North America. Landfills pose the threat of pollution, and can intoxicate ground water. The signs of pollution are effectively masked by disposal companies and it is often hard to see any evidence. Usually landfills are surrounded by large walls or fences hiding the mounds of debris. Large amounts of chemical odor eliminating agent are sprayed in the air surrounding landfills to hide the evidence of the rotting waste inside the plant.

Energy Generation

Municipal solid waste can be used to generate energy. Several technologies have been developed that make the processing of MSW for energy generation cleaner and more economical than ever before, including landfill gas capture, combustion, pyrolysis, gasification, and plasma arc gasification. While older waste incineration plants emitted high levels of pollutants, recent regulatory changes and new technologies have significantly reduced this concern. United States Environmental Protection Agency (EPA) regulations in 1995 and 2000 under the Clean Air Act have succeeded in reducing emissions of dioxins from waste-to-energy facilities by more than 99 percent below 1990 levels, while mercury emissions have been reduced by over 90 percent. The EPA noted these improvements in 2003, citing waste-to-energy as a power source "with less environmental impact than almost any other source of electricity".

Chemical Waste

Chemical waste is a waste that is made from harmful chemicals (mostly produced by large factories). Chemical waste may fall under regulations such as COSHH in the United Kingdom, or the Clean Water Act and Resource Conservation and Recovery Act in the United States. In the U.S., the Environmental Protection Agency (EPA) and the Occupational Safety and Health Administration (OSHA), as well as state and local regulations also regulate chemical use and disposal. Chemical waste may or may not be classed as hazardous waste. A chemical hazardous waste is a solid, liquid, or gaseous material that displays either a "Hazardous Characteristic" or is specifically "listed" by name as a hazardous waste. There are four characteristics chemical wastes may have to be considered as hazardous. These are Ignitability, Corrosivity, Reactivity, and Toxicity. This type of hazardous waste must be categorized as to its identity, constituents, and hazards so that it may be safely handled and managed. Chemical waste is a broad term and encompasses many types of materials. Consult the Material Safety Data Sheet (MSDS), Product Data Sheet or Label for a list of constituents. These sources should state weather this chemical waste is a waste that needs special disposal.

Chemical Waste Bin (Chemobox)

Guidance for Disposal of Laboratory Chemical Wastes

In the laboratory, chemical wastes are usually segregated on-site into appropriate waste carboys, and disposed by a specialist contractor in order to meet safety, health, and legislative requirements.

Innocuous aqueous waste (such as solutions of sodium chloride) may be poured down the sink. Some chemicals are washed down with excess water. This includes: concentrated and dilute acids and alkalis, harmless soluble inorganic salts (all drying agents), alcohols containing salts, hypochlorite solutions, fine (tlc grade) silica and alumina. Aqueous waste containing toxic compounds are collected separately

Chemical Waste Disposal Guideline			
Innocuous aqueous waste	Organic Solvent	Red List	Solid Waste
• Acid (pH<4) • Alkali (pH>10) • Harmless soluble inorganic salt • Alcohol containing salt • Hypochlorite solution • Fine (tlc grade) silica and alumina These chemicals should be washed down with	• **Chlorinated** Example: DCM, Chloroform, Chlorobenzene etc. • **Non-Cholronated** Example: THF, ethyl acetate, hexane, toluene, methanol, etc	• Compounds with transitional metals • Biocides • Cyanides • Mineral oils and hydrocarbons • Poisonous organosilicon compounds • Metal phosphides • Phosphorus element • Fluorides and nitrites.	• Lightly contaminated Example: Gloves, empty vials/centrifuge . Broken Glassware Broken glassware are usually collected in plastic lined cardboard boxes to landfilling. Due in contamination, they are usually not suitable for recycling.

Chemical waste category that should be followed for proper packaging, labelling, and disposal of chemical waste.

Waste elemental mercury, spent acids and bases may be collected separately for recycling.

Waste organic solvents are separated into chlorinated and non-chlorinated solvent waste. Chlorinated solvent waste is usually incinerated at high temperature to minimize the formation of dioxins. Non-chlorinated solvent waste can be burned for energy recovery.

In contrast to this, chemical materials on the "Red List" should never be washed down a drain. This list includes: compounds with transitional metals, biocides, cyanides, mineral oils and hydrocarbons, poisonous organosilicon compounds, metal phosphides, phosphorus element, and fluorides and nitrites.

Moreover, the Environmental Protection Agency (EPA) prohibits disposing certain materials down any UVM drain. Including flammable liquids, liquids capable of causing damage to wastewater facilities (this can be determined by the pH), highly viscous materials capable of causing an obstruction in the wastewater system, radioactive materials, materials that have or create a strong odor, wastewater capable of significantly raising the temperature of the system, and pharmaceuticals or endocrine disruptors.

Broken glassware are usually collected in plastic-lined cardboard boxes for landfilling. Due to contamination, they are usually not suitable for recycling. Similarly, used hypodermic needles are collected as sharps and are incinerated as medical waste.

Chemical Compatibility Guideline

Many chemicals may react adversely when combined. It's recommended that incompatible chemicals are stored in separate areas of the lab.

Acids should be separated from alkalis, metals, cyanides, sulfides, azides, phosphides, and oxidizers. The reason being, when combined acids with these type of compounds, violent exothermic reaction can occur possibly causing flammable gas, and in some cases explosions.

Oxidizers should be separated from acids, organic materials, metals, reducing agents, and ammonia. This is because when combined oxidizers with these type of compounds, inflammable, and sometimes toxic compounds can occur.

Container Compatibility

When disposing hazardous laboratory chemical waste, chemical compatibility must be considered. For safe disposal, the container must be chemically compatible with the material it will hold. Chemicals must not react with, weaken, or dissolve the container or lid. Acids or bases should not be stored in metal. Hydrofluoric acid should not store in glass. Gasoline (solvents) should not store or transport in lightweight polyethylene containers such as milk jugs. Moreover, the Chemical Compatibility Guidelines should be considered for more detailed information.

Laboratory Waste Containers

Packaging, labelling, storage are the three requirements for disposing chemical waste.

Packaging

For packaging, chemical liquid waste containers should only be filled up to 75% capacity to allow for vapour expansion and to reduce potential spills which could occur from moving overfilled containers. Container material must be compatible with the stored hazardous waste. Finally, wastes must not be packaged in containers that improperly identify other nonexisting hazards.

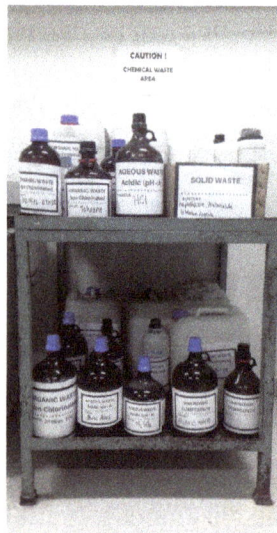

How to properly label, package, and store chemical waste safely.

In addition to the general packaging requirements mentioned above, incompatible materials should never be mixed together in a single container. Wastes must be stored in containers compatible with the chemicals stored as mentioned in the container compatibility section. Solvent safety cans should to be used to collect and temporarily store large volumes (10-20 litres) of flammable organic waste solvents, precipitates, solids or other non-fluid wastes should not be mixed into safety cans.

Labelling

Label all containers with the group name from the chemical waste category and an itemized list of the contents. All chemicals or anything contaminated with chemicals posing a significant hazard. All waste must be appropriately packaged.

Storage

When storing chemical wastes, the containers must be in good condition and should remain closed unless waste is being added. Hazardous waste must be stored safely prior to removal from the laboratory and should not be allowed to accumulate. Container should be sturdy and leakproof, also has to be labeled. All liquid waste must be stored in leakproof containers with a screw- top or other secure lid. Snap caps, mis-sized caps, parafilm and other loose fitting lids are not acceptable. If necessary, transfer waste material to a container that can be securely closed. Keep waste containers closed except when adding waste. Secondary containment should be in place to capture spills and leaks from the primary container, segregate incompatible hazardous wastes, such as acids and bases.

Mapping of Chemical Waste in the United States

TOXMAP is a Geographic Information System (GIS) from the Division of Specialized Information Services of the United States National Library of Medicine (NLM) that uses maps of the United States to help users visually explore data from the United States Environmental Protection Agency's (EPA) Toxics Release Inventory and Superfund Basic Research Programs. TOXMAP is a resource funded by the US Federal Government. TOXMAP's chemical and environmental health information is taken from NLM's Toxicology Data Network (TOXNET) and PubMed, and from other authoritative sources.

Chemical waste in Canadian Aquaculture

Chemical waste in our oceans is becoming a major issue for the marine life. There have been many studies conducted to try and prove the effects of these chemical in our oceans. In Canada, many of the studies concentrated on the Atlantic provinces, where fishing and aquaculture are an important part of the economy. In New Brunswick, a study was done on the sea urchin in an attempt to identify the effects of toxic and chemical waste on life beneath the ocean, specifically the wasted from the salmon farms. Sea urchins were used to check the levels of metals in the environment. It is advantageous to use green sea urchins, Strongylocentrotus droebachiensis, because they are widely distributed, abundant in many locations, and easily accessible. By investigating the concentrations of metals in the green sea urchins, the impacts of produced chemicals from salmon aquaculture activity could be assessed and detected. Samples were taken at 25m intervals along a transect in the direction of the main tidal flow. The study found that there was impacts to at least 75m based on the intestine metal concentrations. So based on this study it is clear that the metals are contaminating the oceans and negatively affecting aquatic life.

Green Sea Urchin or S. droebacheinsis

Uranium in Ground and Surface Water in Canada

Another issue regarding chemical waste is the potential risk of surface and groundwater contamination by the heavy metals and radionuclides leached from uranium waste-rock piles (UWRP) A Radionuclide is an atom that has excess nuclear energy, making it unstable. Uranium waste-rock piles refers to Uranium mining, which is the process of extraction of uranium ore from the ground. . An example of such threats is in Saskatchewan, Uranium mining and ore processing (milling) can pose a threat to the environment. In open pit mining, large amounts of materials are excavated and disposed off in waste-rock piles. Waste-rock piles from the Uranium mining industry can contain several heavy metals and contaminants that may become mobile under certain conditions. Environmental contaminants may include acid mine drainage, higher concentrations of radionuclides, and non-radioactive metals/metalloids (i.e. As, Mo, Ni, Cu, Zn).

The leachability of heavy metals and radionuclide from UWRP plays a significant role in determining their potential environmental risks to surrounding surface and groundwater. Substantial differences in the solid-phase partitioning and chemical leachability of Ni and U were observed in the investigated UWRP lithological materials and background organic-rich lake sediment. For Instance, in the uranium-mining district of Northern Saskatchewan, Canada, the sequential extraction results showed that a significant amount of Ni (Nickel) was present in the non-labile residual fraction, while Uranium was mostly distributed in the moderately labile fractions. Although Nickel was much less labile than Uranium, the observed Nickel exceeded Uranium concentrations in leaching]].The observed Nickel and Uranium concentrations were relatively high in the underlying organic-rich lake sediment. Expressed as the percentage of total metal content, potential leachability decreased in the order U > Ni. Data suggest that these elements could potentially migrate to the water table below the UWRP. Detailed information regarding the solid-phase distribution of contaminants in the UWRP is critical to understand the potential for their environmental transport and mobility

Image of Uranium risk map may be found here: http://www.env.gov.nl.ca/env/waterres/cycle/groundwater/well/uranium.pdf

The most visible civilian use of uranium is as the thermal power source used in nuclear power plants

Biomedical Waste

Biomedical waste is potentially infectious. Biomedical waste may also include waste associated with the generation of biomedical waste that visually appears to be of medical or laboratory origin (e.g., packaging, unused bandages, infusion kits, etc.), as well research laboratory waste containing biomolecules or organisms that are restricted from environmental release. As detailed below, discarded sharps are considered biomedical waste whether they are contaminated or not, due to the possibility of being contaminated with blood and their propensity to cause injury when not properly contained and disposed of. Biomedical waste is a type of biowaste.

Biomedical waste in containers, held in accumulation area awaiting treatment

Biomedical waste may be solid or liquid. Examples of infectious waste include discarded blood, sharps, unwanted microbiological cultures and stocks, identifiable body parts (including those as a result of amputation), other human or animal tissue, used bandages and dressings, discarded gloves, other medical supplies that may have been in contact with blood and body fluids, and laboratory waste that exhibits the characteristics described above. Waste sharps include potentially contaminated used (and unused discarded) needles, scalpels, lancets and other devices capable of penetrating skin.

Biomedical waste is generated from biological and medical sources and activities, such as the diagnosis, prevention, or treatment of diseases. Common generators (or producers) of biomedical waste include hospitals, health clinics, nursing homes, medical research laboratories, offices of physicians, dentists, and veterinarians, home health care, and funeral homes. In healthcare facilities (i.e., hospitals, clinics, doctors offices, veterinary hospitals and clinical laboratories), waste with these characteristics may alternatively be called medical or clinical waste.

Biomedical waste is distinct from normal trash or general waste, and differs from other types of hazardous waste, such as chemical, radioactive, universal or industrial waste. Medical facilities generate waste hazardous chemicals and radioactive materials. While such wastes are normally not infectious, they require proper disposal. Some wastes are considered *multihazardous,* such as tissue samples preserved in formalin.

Risk to Human Health

Disposal of this waste is an environmental concern, as many medical wastes are classified as *infectious* or *biohazardous* and could potentially lead to the spread of infectious disease. The most common danger for humans is the infection which also affects other living organisms in the region. Daily exposure to the waste (landfill) leads to accumulation of harmful substances or microbes in the person's body.

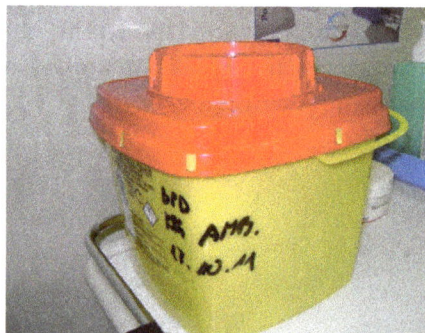
"Sharps container" for needles

A 1990 report by the U.S. Agency for Toxic Substances and Disease Registry concluded that the general public is not likely to be adversely affected by biomedical waste generated in the traditional healthcare setting. They found, however, that biomedical waste from those settings may pose an injury and exposure risks via occupational contact with medical waste for doctors, nurses, and janitorial, laundry and refuse workers. Further, there are opportunities for the general public to come into contact medical waste, such as needles used illicitly outside healthcare settings, or biomedical waste generated via home health care.

Management

Biomedical waste must be properly managed and disposed of to protect the environment, general public and workers, especially healthcare and sanitation workers who are at risk of exposure to biomedical waste as an occupational hazard. Steps in the management of biomedical waste include generation, accumulation, handling, storage, treatment, transport and disposal.

On-site Versus Off-site

Disposal occurs off-site, at a location that is different from the site of generation. Treatment may occur on-site or off-site. On-site treatment of large quantities of biomedical waste usually requires the use of relatively expensive equipment, and is generally only cost effective for very large hospitals and major universities who have the space, labor and budget to operate such equipment. Off-site treatment and disposal involves hiring of a biomedical waste disposal service (also called a truck service) whose employees are trained to collect and haul away biomedical waste in special containers (usually cardboard boxes, or reusable plastic bins) for treatment at a facility designed to handle biomedical waste.

These healthcare workers are being trained to safely handle contaminated wastes before being assigned to an outbreak of Ebola hemorrhagic fever.

Generation and Accumulation

Biomedical waste should be collected in containers that are leak-proof and sufficiently strong to prevent breakage during handling. Containers of biomedical waste are marked with a biohazard symbol. The container, marking, and labels are often red.

Discarded sharps are usually collected in specialized boxes, often called *needle boxes*.

Specialized equipment is required to meet OSHA 29 CFR 1910.1450 and EPA 40 CFR 264.173. standards of safety. Minimal recommended equipment include a fume hood and primary and secondary waste containers to capture potential overflow. Even beneath the fume hood, containers containing chemical contaminants should remain closed when not in use. An open funnel placed in the mouth of a waste container has been shown to allow significant evaporation of chemicals into the surrounding atmosphere, which is then inhaled by laboratory personnel, and contributes a primary component to the threat of completing the fire triangle. To protect the health and safety of laboratory staff as well as neighboring civilians and the environment, proper waste management equipment, such as the Burkle funnel in Europe and the ECO Funnel in the U.S., should be utilized in any department which deals with chemical waste. It is to be dumped after treatment.

Handling

Handling refers to the act of manually moving biomedical waste between the point of generation, accumulation areas, storage locations and on-site treatment facilities. Workers who handle biomedical waste should observe *standard precautions*.

Treatment

The goals of biomedical waste treatment are to reduce or eliminate the waste's hazards, and usually to make the waste unrecognizable. Treatment should render the waste safe for subsequent handling and disposal. There are several treatment methods that can accomplish these goals.

Biomedical waste is often incinerated. An efficient incinerator will destroy pathogens and sharps. Source materials are not recognizable in the resulting ash.

An autoclave may also be used to treat biomedical waste. An autoclave uses steam and pressure to sterilize the waste or reduce its microbiological load to a level at which it may be safely disposed of. Many healthcare facilities routinely use an autoclave to sterilize medical supplies. If the same autoclave is used to sterilize supplies and treat biomedical waste, administrative controls must be used to prevent the waste operations from contaminating the supplies. Effective administrative controls include operator training, strict procedures, and separate times and space for processing biomedical waste.

For liquids and small quantities, a 1–10% solution of bleach can be used to disinfect biomedical waste. Solutions of sodium hydroxide and other chemical disinfectants may also be used, depending on the waste's characteristics. Other treatment methods include heat, alkaline digesters and the use of microwaves.

For autoclaves and microwave systems, a shredder may be used as a final treatment step to render the waste unrecognizable.

Country-wise Regulation and Management

The international symbol for biological hazard.

United Kingdom

In the UK, clinical waste and the way it is to be handled is closely regulated. Applicable legislation includes the Environmental Protection Act 1990 (Part II), Waste Management Licensing Regulations 1994, and the Hazardous Waste Regulations (England & Wales) 2005, as well as the Special Waste Regulations in Scotland.

United States

In the United States, biomedical waste is usually regulated as medical waste. In 1988 the U.S. federal government passed The Medical Waste Tracking Act which set the standards for governmental regulation of medical waste. After the Act expired in 1991, States were given the responsibility to regulate and pass laws concerning the disposal of medical waste. All fifty states vary in their regulations from no regulations to very strict.

In addition to on-site treatment or pickup by a biomedical waste disposal firm for off-site treatment, a mail-back disposal option exists in the United States. In mail-back biomedical waste disposal, the waste is shipped through the U.S. postal service instead of transport by private hauler. While currently available in all 50 U.S. states, mail-back medical waste disposal is limited to very strict postal regulations (i.e., collection and shipping containers must be approved by the postal service for use) and only available by a handful of companies.

India

In India, The Bio-medical Waste (Management and Handling) Rules, 1998 and further amendments were passed for the regulation of bio-medical waste management. On 28 th Mar 2016 Biomedical Waste Management Rules 2016 were also notified by Central Govt. Each state's Pollution Control Board or Pollution control Committee will be responsible for implementing the new legislation.

In India, there are a number of different disposal methods, yet most are harmful rather than helpful. If body fluids are present, the material needs to be incinerated or put into an autoclave. Although this is the proper method, most medical facilities fail to follow the regulations. It is often found that biomedical waste is put into the ocean, where it eventually washes up on shore, or in landfills due to improper sorting when in the medical facility. Improper disposal can lead to many diseases in animals as well as humans. For example, animals, such as cows in Pondicherry, India, are consuming the infected waste and eventually, these infections can be transported to humans through eating of the meat.

Many studies took place in Gujarat, India regarding the knowledge of workers in facilities such as hospitals, nursing homes, or home health. It was found that 26% of doctors and 43% of paramedical staff were unaware of the risks related to biomedical wastes. After extensively looking at the different facilities, many were undeveloped in the area regarding biomedical waste. The rules and regulations in India work with The Bio-medical Waste (Management and Handling) Rules from 1998, yet a large number of health care facilities were found to be sorting the waste incorrectly. Worldwide, there are specific colored bags, bins and labels that are recommended for each type of waste. For example, syringes, needles and blood-soiled bandages should be all disposed of in a red colored bag or bin, where it will later be incinerated.

Commercial Waste

Commercial waste consists of waste from premises used wholly or mainly for the purposes of a trade or business or for the purpose of sport, recreation, education or entertainment, but excluding household, agricultural or industrial waste.

Biodegradable Waste

Biodegradable waste includes any organic matter in waste which can be broken down into carbon dioxide, water, methane or simple organic molecules by micro-organisms and other living things using composting, aerobic digestion, anaerobic digestion or similar processes. In waste management, it also includes some inorganic materials which can be decomposed by bacteria. Such materials include gypsum and its products such as plasterboard and other simple organic sulfates which can be decomposed to yield hydrogen sulfide in anaerobic land-fill conditions.

In domestic waste collection, the scope of biodegradable waste may be narrowed to include only those degradable wastes capable of being handled in the local waste handling facilities.

Sources

Biodegradable waste can be commonly found in municipal solid waste (sometimes called biodegradable municipal waste, or BMW) as green waste, food waste, paper waste, and biodegradable plastics. Other biodegradable wastes include human waste, manure, sewage, sewage sludge and slaughterhouse waste. In the absence of oxygen, much of this waste will decay to methane by anaerobic digestion.

In many parts of the developed world, biodegradable waste is separated from the rest of the waste stream, either by separate kerb-side collection or by waste sorting after collection. At the point of collection such waste is often referred to as *Green waste*. Removing such waste from the rest of the waste stream substantially reduces waste volumes for disposal and also allows biodegradable waste to be composted where composting facilities exist.

Uses of Biodegradable Waste

Biodegradable waste can be used for composting or a resource for heat, electricity and fuel by means of incineration or anaerobic digestion. Swiss *Kompogas* and the Danish *AIKAN* process are examples of anaerobic digestion of biodegradable waste. While incineration can recover the most energy, anaerobic digestion plants retain the nutrients and compost for the soil and still recover some of the contained energy in the form of biogas. Kompogas produced 27 million Kwh of electricity and biogas in 2009. The oldest of the company's own lorries has achieved 1,000,000 kilometers driven with biogas from household waste in the last 15 years.

Areas Relying on Organic Waste

Featured in an edition of *The Economist* that predicted events in 2014, it was revealed that Massachusetts creates roughly 1.4 million tons of organic waste every year. Massachusetts, along with Connecticut and Vermont, are also going to enact laws to divert food waste from landfills.

In small and densely populated states, landfill capacity is limited so disposal costs are higher ($60–90 per ton in MA compared to national average of $45). Decomposing food waste generates methane, a notorious greenhouse gas. However, this biogas can be captured and turned into energy through anaerobic digestion, and then sold into the electricity grid.

Anaerobic digestion grew in Europe, but is starting to develop in America. Massachusetts is increasing its production of anaerobic digesters.

Climate Change Impacts

The main environmental threat from biodegradable waste is the production of methane and other greenhouse gases.

References

- Hallam, Bill (April–May 2010). "Techniques for Efficient Hazardous Chemicals Handling and Disposal". Pollution Equipment News. p. 13. Retrieved 10 March 2016.

- "LABORATORY CHEMICAL WASTE MANAGEMENT GUIDELINES" (PDF). Environmental Health and Radiation Safety University of Pennsylvania. Retrieved 10 March 2016.

- "Waste - Disposal of Laboratory Wastes (GUIDANCE) | Current Staff | University of St Andrews". www.st-andrews.ac.uk. Retrieved 2016-02-04.

- Laboratory, National Research Council (US) Committee on Prudent Practices in the (2011-01-01). "Management of Waste". Retrieved 10 March 2016.

- "PROCEDURES FOR LABORATORY CHEMICAL WASTE DISPOSAL" (PDF). Memorial University. Retrieved 10 March 2016.

- "Crude tall oil feed stocks cannot be considered 'waste'". Moran, Kevin, Financial Times. April 30, 2014. Retrieved 2014-07-03.

- "National Research Council Recommendations Concerning Chemical Hygiene in Laboratories". United States Department of Labor. Retrieved 15 May 2013.

3

Waste Handling Practices

Waste needs to be handled in order to keep our environment and our surroundings clean. There are a number of practices that handle waste; some of these are kerbside collection, garbage truck, sand cleaning machine and waste collection. The service provided to urban cities for removing waste is termed as kerbside collection. The chapter strategically encompasses and incorporates the major components and key concepts of waste management, providing a complete understanding.

Kerbside Collection

Kerbside collection, or curbside collection, is a service provided to households, typically in urban and suburban areas of removing household waste. It is usually accomplished by personnel using purpose built vehicles to pick up household waste in containers acceptable to or prescribed by the municipality.

Kerbside collection in Canberra, Australia

History

Prior to the 20th century the amount of waste generated by a household was relatively small. Household wastes were often simply thrown out the window, buried in the garden or deposited in outhouses. When human concen-trations became more dense, waste collectors, called nightmen or gong farmers were hired to collect the night soil from pail closets, performing their duties only at night (hence the name). Meanwhile, disposing of refuse became a problem wherever cities grew. Often refuse was placed in unusable areas just outside the city, such as wetlands and tidal zones. One example is London, which from Roman times disposed of its refuse outside the London Wall beside the River Thames. Another example is 1830s Manhattan, where thousands of hogs were permitted to roam the streets and eat garbage. A small industry developed as "swill children" collected kitchen refuse to sell for pig feed and the rag and bone man traded goods for bones (used for glue) and rags (essential for paper manufacture prior to the invention of wood pulping). Later, in the late nineteenth century, trash was fed to swine in industrial.

As sanitation engineering came to be practised beginning in the mid-19th century and human waste was conveyed from the home in pipes, the gong farmer was replaced by the municipal trash collector as there remained growing amounts of household refuse, including fly ash from coal, which was burnt for home heating. In Paris, the rag and bone man worked side by side with the municipal bin man, though reluctantly: in 1884, Eugène Poubelle introduced the first integrated kerbside collection and recycling system, requiring residents to separate their waste into perishable items, paper and cloth, and crockery and shells. He also established rules for how private collectors and city workers should cooperate and he developed standard dimensions for refuse containers: his name in France is now synonymous with the garbage can. Under Poubelle, food waste and other organics collected in Paris were transported to nearby Saint Ouen where they were composted. This continued well into the 20th century when plastics began to contaminate the waste stream.

From the late-19th century to the mid-20th century, more or less consistent with the rise of consumables and disposable products municipalities began to pass anti-dumping ordinances and introduce kerbside collection. Residents were required to use a variety of refuse containers to facilitate kerbside collection but the main type was a variation of Poubelle's metal garbage container. It was not until the late 1960s that the green bin bag was introduced by Glad. Later, as waste management practices were introduced with the aim of reducing landfill impacts, a range of container types, mostly made of durable plastic, came to be introduced to facilitate the proper diversion of the waste stream. Such containers include blue boxes, green bins and wheelie bins or MGBs.

Over time, waste collection vehicles gradually increased in size from the hand pushed tip cart or English dust cart, a name by which these vehicles are still referred, to large compactor trucks.

Waste Management and Resource Recovery

Kerbside collection is today often referred to as a strategy of local authorities to collect recyclable items from the consumer. Kerbside collection is considered a low-risk strategy to reduce waste volumes and increase recycling rates. Materials are typically collected in large bins, coloured bags, or small open plastic tubs, specifically designated for content.

Glass for collection in Edinburgh, Scotland.

Recyclable materials that may be separately collected from municipal waste include:

Biodegradable waste component

- Green waste

- Kitchen waste

Recyclable materials, depending on location

- Office paper

- Newsprint

- Paperboard

- Corrugated fiberboard

- Plastics (#1 PET, #2 HDPE natural and colored, #3 PVC narrow-necked containers, #4 LDPE, #5 PP, #6 Polystyrene (however not EXPANDED polystyrene, an example of recyclable polystyrene may be a yoghurt pot) #7 other mixed resin plastics)

- Glass

- Copper

- Aluminum

- Steel and Tinplate

- Co-mingled recyclables- can be sorted by a clean materials recovery facility

Kerbside collection of recyclable resources is aimed to recover purer waste streams with higher market value than by other collection methods. If the household incorrectly separates the recyclable elements, the load may have to be put to landfill if it is deemed to be contaminated.

Kerbside collection and household recycling schemes are also being used as tools by local authorities to increase the public's awareness of their waste production.

In Somerville, MA all accepted paper, glass, plastic, and metal recycling is picked up from a single bin

Kerbside collection is commonly considered to be completely environmentally friendly. This may not necessarily be the case as it leads to an increased number of waste collection vehicles on the road, in themselves contributing to global warming through exhaust emissions until the time of their conversion to clean energy.

New and emerging waste treatment technologies such as mechanical biological treatment may offer an alternative to kerbside collection through automated separation of waste in recycling factories.

Usage by Country

Canada

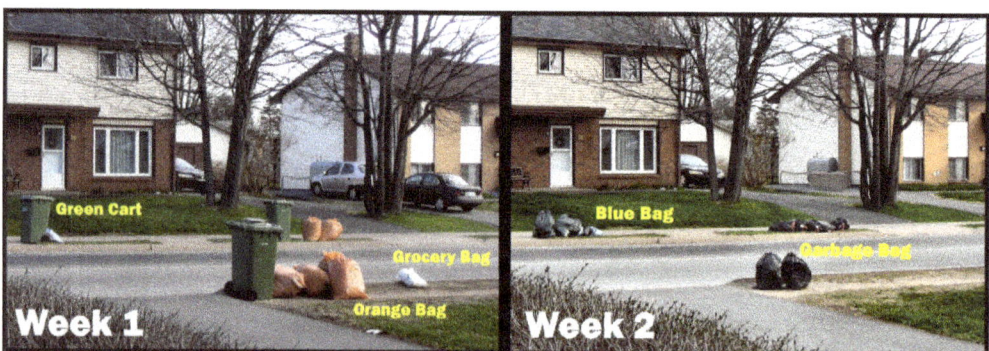

Halifax Regional Municipality (HRM) in Nova Scotia, Canada, with a population of about 375,000, has one of the most complex kerbside collection programmes in North America. Based on the green cart, it requires residents to self-sort refuse and place different types at the kerb on alternating weeks. As shown in the photo at left, week 1 would see the green cart and optional orange bags used for kitchen waste and other organics such as yard waste. Week 2 would permit non-recoverable waste in garbage bags or cans. Blue bags are used for paper, plastic and metal containers. Together with used grocery bags containing newspapers, they may be placed on the kerb either week. In summer, the green cart is emptied weekly due to the prevalence of flies. HRM has achieved a diversion rate of approximately 60 percent by this method.

Calgary, Alberta has adopted "Curbside" Recycling and uses blue bins. The blue cart programme accepts all types of recyclables, including plastics 1-7. It is picked up weekly for the cost of $8.00 per month. This programme is mandatory.

In 1981 Resource Integration Systems (RIS) in collaboration with Laidlaw International tested the first blue box recycling system on 1500 homes in Kitchener, Ontario. Due to the success of the project the City of Kitchener put out a contract for public bid in 1984 for a recycling system city wide. Laidlaw won the bid and continued with the popular blue box recycling system. Today hundreds of cities around the world use the blue box system or a similar variation.

Many Canadian municipalities use "green bins" for kerbside recycling. Others, such as Moncton, use wet/dry waste separation and recovery programmes.

New Zealand

In New Zealand, kerbside collection of general refuse and recycling, and in some areas organic waste, is the responsibility of the local city or district council, or private contractors. Practices and collection methods vary widely from council to council and company to company. Some examples of collection are:

Kerbside collection bins in Dunedin, New Zealand. The yellow-liddied wheelie bin is for non-glass recyclables, and the blue bin is for glass. The two bins are collected on alternating weeks. Official council bags are used for general household waste, and are collected weekly.

- Auckland City Council: Two 240-litre wheelie bins are supplied: a red-lidded bin for general refuse, collected weekly, and a blue-lidded bin for recyclables, collected fortnightly.

- Christchurch City Council: Three wheelie bins are supplied: a 140-litre red-lidded bin for general refuse, a 240-litre yellow-lidded bin for recyclables, and an 80-litre green-lidded bin for organic waste. The organic waste bins are collected weekly, while the recyclables and general refuse bins are collected on alternating weeks.

- Hamilton City Council and Hutt City Council: A 45-litre bin is supplies for recyclables, collected weekly. General refuse is collected weekly using user-pays official council bags.

- Dunedin City Council, Palmerston North City Council and Wellington City Council: Two bins are supplied: a 45-litre or 70-litre bin for glass, and an 80-litre or 240-litre wheelie bin for non-glass recyclables. These two bins are collected on alternating weeks. General refuse is collected weekly using user-pays official council bags.

- Rodney District Council: A 45-litre bin is supplies for recyclables, collected weekly. There is no council collection of general waste, and all general waste collection is carried out by independent companies.

- Taupo District Council: A 45-litre bin is supplies for recyclables, collected weekly. General refuse is collected weekly using user-pays system of orange tags - one orange tag is to be placed on a standard rubbish bag up to 60 litres capacity, or half an orange sticker can be placed on two supermarket bags tied together.

- Upper Hutt City Council: Recycling is to be placed in plastic bags, with paper and cardboard collected in the first week, and plastic, metal and glass in the second week. General refuse is collected weekly using user-pays official council bags.

- Waitakere City Council: A 140-litre wheelie bin is provided for recyclables, collected fortnightly. General refuse is collected weekly using user-pays official council bags.

By 1996 the New Zealand cities of Auckland, Waitakere, North Shore and Lower Hutt had kerbside recycling bins available. In New Plymouth, Wanganui and Upper Hutt recyclable material was collected if placed in suitable bags. By 2007 73% of New Zealanders had access to kerbside recycling.

Kerbside collection of organic waste is carried out by the Mackenzie District Council and the Timaru District Council. Christchurch City Council is introducing the system to their kerbside collection. Other councils are carrying out trials.

United Kingdom

In the United Kingdom, the Household Waste Recycling Act 2003 requires local authorities to provide every household with a separate collection of at least two types of recyclable materials by 2010. There has been criticism in the difference of schemes used in the country such as the colour of bins, whether they are bins boxes or bags, and also the fact that clutter roads and how the additional trucks and collections needed have carbon dioxide emissions too. Some find the colour differences confusing, and people want a national scheme. A typical example is to compare two neighbouring councils in greater Manchester, Bury council and Salford. Bury uses blue for cans, plastic and glass, green for paper and cardboard and brown for garden waste. Salford uses blue for paper and card, brown for cans plastic and glass and pink for garden waste. Most councils use grey or black for general waste, with a few exceptions such as Liverpool, which uses purple for general waste, a colour used by no other council

Another controversial issue in the uk is the frequency of the waste collections. To save money, many councils are cutting the frequency of both general waste and recyclables collections. This has led to problems from larger families, and has led to overflowing and fly tipping. For example, previously, Bury Council collected general waste once a week and recyclables fortnightly. This has now changed to fortnightly for general waste and monthly (every 4 weeks) collection of recyclables.

A few councils are using "forced" recycling, by replacing the large, 240l general waste bin with a smaller 180l or 140l bin, and using the old 240l one for recyclables. This may be made worse by fortnightly collections of the "small" bin, and strict rules such as "No extra bags will be taken" and "Bin lids must be fully closed". Stockport Council is a notable user of this scheme. Their recycling rates have risen substantially as a result, but there are usually complaints from families. Trafford council also use a similar scheme, but the small grey bin is emptied every week. In addition, the two named councils, and more, collect food waste together with garden waste, by sending out kitchen caddies and compostable liners. These prevent food waste (including meat) from going to landfill, and to increase the councils recycling rate. The food and garden waste is usually collected weekly or fortnightly, and is taken to an In Vessel composter or Anaerobic digester, where the waste is turned into soil improver for use on local farms.

In the north west, all the glass collected is used within the UK, around half of the plastics and cans are used in the UK; the rest is sent further afield to Europe or China to be made into new products, and paper and paperboard collected is sent to local paper mills to be made into newspapers, tissues, paperboard and office paper. Again some of the paper will be sent further afield.

Some councils only have 3 bins- general, organic and recyclables. This means that plastics, cans and glass go in the same container as paper and cardboard. Although this is much easier for the residents, there is more sorting required, and the paper quality is sometimes of a low grade due to food contamination or shards of glass in the paper, and so this scheme is criticized.

Also, most councils require residents to remove caps from bottles and rinse them out to avoid smells. This is because the lids are made from a different type of plastic (PP) to the bottle (PET/HDPE) - although by collapsing the bottles and folding them over like toothpaste tubes and rescrewing the caps in place enables the volume of bottles to be drastically reduced, thereby increasing the amount of bottles that can be carried in the recycling bins. In fact many bottlers, especially bottled water companies, have now designed their bottles to be collapsible; though this message has not been effectively disseminated to the consumer. A collapsable bottle takes between 25% and 33% of the space a non-collapsed bottle.

Labels are rarely required to be removed, however. This also means that only plastic bottles are recycled. Councils are still trying to make clear that plastic tubs (yogurts, desserts and spreads), bags and cling film cannot be recycled through the kerbside economically. If too much contamination is collected then this results in the whole vehicle load going to landfill at a high cost. Contamination is usually a problem if recyclables are collected in wheelie bins, as the worker can only look at the top; there may be contamination 'hidden' at the bottom. Councils that use many bags and boxes (Edinburgh) suffer from less contamination but are complicated and the loose paper and cardboard, and recycling bags are blown around, and paper can be wet.

Basque Country

In the province of Gipuzkoa, this system is implanted in many towns as Usurbil, Hernani, Oiartzun, Antzuola, Legorreta, Itsasondo, Zaldibia, Anoeta, Alegia, Irura, Zizurkil, Astigarraga, Ordizia, Oñati and Lezo, where the common used name in basque is atez-atekoa, which means *door-by-door*. Due to the big success in this towns, with more than the 80% of the waste recycled, 34 towns in Gipuzkoa are studying to set this system up in 2013, like Arrasate, Bergara, Aretxabaleta, Eskoriatza, Legazpi, Tolosa or Pasaia.

The atez-ate system consists in hanging each kind of rubbish in a hanger outside the house a certain day or days in a week. For example, in Hernani, they have three days to hang their organic rubbish, two days for plastics and metallics, one for paper and one for rejects residuals.

This system started in the town of Usurbil in the year 2009, due to the incinerator of the region of Gipuzkoa was going to build in this town, exactly in the neighborhood of Zubieta. Three years after, the construction of the incinerator was paralyzed by the government of the region, suggesting that the incinerator was a source of contamination and the high cost of the building.

Criticism

This type of collection service is subject to growing criticism.

- The large (Wheelie bin) container encourages the "out of sight" rubbish mentality and invites more rubbish to be disposed.

- The bins and collection trucks are not suited to narrow roads or houses with steep driveways or steps.

- They lock local authorities into capital intensive equipment programmes and multi-national providers.

- Co-mingled recyclables are sometimes not being successfully managed by automated sorting stations and the rates of diversion are low. In some cases, this results in mountains of unsorted recyclables.

- In the UK especially, some councils are sending out at least 4 large bins - residents of smaller houses with no gardens have little space to put them

- Many use small plastic boxes, bags and lockable outdoor food waste 'caddies' which get blown around and lost, bad for recycling participation.

Automated Vacuum Collection

An **a**utomated vacuum waste collection system, also known as pneumatic refuse collection, or automated vacuum collection (AVAC), transports waste at high speed through underground pneumatic tubes to a collection station where it is compacted and sealed in containers. When the container is full, it is transported away and emptied. The system helps facilitate separation and recycling of waste.

The process begins with the deposit of trash into intake hatches, called portholes, which may be specialized for waste, recycling, or compost. Portholes are located in public areas and on private property where the owner has opted in. The waste is then pulled through an underground pipeline by air pressure difference created by large industrial fans, in response to porthole sensors that indicate when the trash needs to be emptied and help ensure that only one kind of waste material is travelling through the pipe at a time. The pipelines converge on a central processing facility that uses automated software to direct the waste to the proper container, from there to be trucked to its final location, such as a landfill or composting plant.

History

The first system was created in Sweden in the 1960s, designed by the Swedish corporation Envac AB (formerly known as Centralsug AB). The first installation was in 1961 at Sollefteå Hospital. The first vacuum system for household waste, was installed in the new residential district of Ör-Hallonbergen, Sweden in 1965.

Current Systems

The Envac proprietary system, Envac Automated Waste Collection System, is used in more than 30 countries.

There are close to a thousand systems in operation all over the world - in China, South East Asia, Korea, the Middle East, the United States, and Europe. In the U.S., this type of system is installed in several places but Disney World and Roosevelt Island are the best known.

Major cities in which the system is operating include Copenhagen, Barcelona, London, and Stockholm.

Pneumatic refuse collection in Vitoria-Gasteiz, Basque Country, Northern Spain

Israel

In February 2007, Yavne municipality issued a tender for the establishment of a pneumatic evacuation of household waste for the residential project "Green Neighborhood", which comprises 4,700 household units. In 2012, the system began operating in residential use. In February 2014, the municipality began replacing the neighborhood's street trash cans with pneumatic cans.

In May 2012, Ra'anana municipality approved the residential project "Neve Zemer" which is planned to include 3,500 housing units and a pneumatic evacuation system of household waste. As of February 2016, the project is expected to complete in 2017.

Planned Systems

Europe

A system is planned to be installed in the new Jätkäsaari residential neighbourhood in Helsinki, Finland. All housing cooperatives and other apartment buildings are obliged to join the network. The system envisioned for Jätkäsaari would help facilitate the separation and recycling of waste.

Each building will have a collection point with up to five wastebins or tubes, each for different types of waste and with the capacity to store several parcels of waste. The underground tube network would act in a manner similar to a packet switched telecommunication network, transporting one kind of waste at a time. Once an input bin is filled, or capacity is available, it is transferred to the central collection site combined with the same class of waste.

Similarly in Finland, a new suburban development, being built in the city of Tampere, will be home to 13,000 inhabitants, creating approximately 5,000 jobs along the way. The suburb of Vuores will have a total of 124 collection points and 368 waste inlets. The system's daily collection capacity for dry waste, bio waste, paper and recyclable cardboard comes to a combined total of 13,000 kilos.

The MetroTaifun Automatic Waste Collection System was selected for Vuores since it consumes only a third of the energy compared to conventional pneumatic waste collection solutions, and half of the traditional garbage truck and container based collection method. When ready, the system will consist of about 400 waste inlets and 13 km of pipe work. The MetroTaifun automatic waste collection system has initially start to collect waste in the 2012.

In Bergen, Norway, a system designed to cover most of the city center is under the first phase of construction. For several years, pipes have been prepared in the ground during other construction projects. The first part of the system is planned to start operations in the fall of 2015, covering about 3000 households.

North America

A system is planned for a new City Center development in Carmel, Indiana. It would service condominiums, businesses, and a hotel.

In March 2015, the city of Montreal abandoned its $3 million investment in a plan to install an automated vacuum collection system in the Quartier des Spectacles entertainment district.

Another installation is planned for Hudson Yards, Manhattan.

Middle East

The world's largest AWCS is now being built in the vicinity of Islam's holiest mosque in Mecca, Saudi Arabia. During the Ramadan and Hajj, 600,000 kilos (or 4,500 cubic meters) of waste is generated each day, which puts a heavy demand on those responsible for collecting the waste and litter. In the MetroTaifun Automatic Waste Collection System, the waste is automatically collected from 74 waste feeding points spread out across the area and then transferred via a 20-kilometre pipe network to a central collection point, keeping all the waste collecting activities out of sight and below ground with the central collection point well away from the public areas.

Mechanical Biological Treatment

A mechanical biological treatment (MBT) system is a type of waste processing facility that combines a sorting facility with a form of biological treatment such as composting or anaerobic digestion. MBT plants are designed to process mixed household waste as well as commercial and industrial wastes.

Process

The terms *mechanical biological treatment* or *mechanical biological pre-treatment* relate to a group of solid waste treatment systems. These systems enable the recovery of materials contained within the mixed waste and facilitate the stabilisation of the biodegradable component of the material.

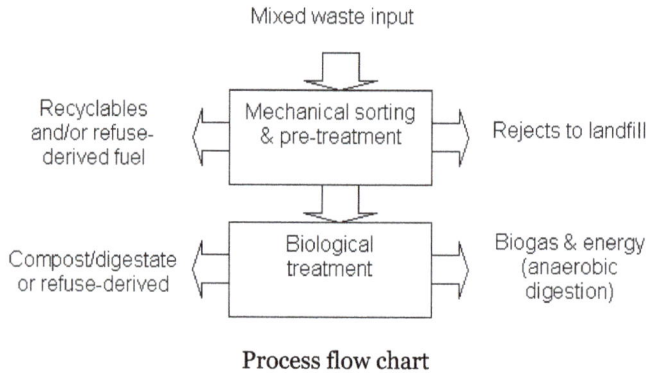

Mixed waste input

Recyclables and/or refuse-derived fuel	Mechanical sorting & pre-treatment	Rejects to landfill
Compost/digestate or refuse-derived	Biological treatment	Biogas & energy (anaerobic digestion)

Process flow chart

The sorting component of the plants typically resemble a materials recovery facility. This component is either configured to recover the individual elements of the waste or produce a Refuse-derived fuel that can be used for the generation of power.

The components of the mixed waste stream that can be recovered include:

- Ferrous Metal

- Non-ferrous metal

- Plastic

- Glass

Terminology

MBT is also sometimes termed BMT – biological mechanical treatment – however this simply refers to the order of processing, i.e. the biological phase of the system precedes the mechanical sorting. MBT should not be confused with MHT – *mechanical heat treatment*

Mechanical Sorting

The "mechanical" element is usually an automated mechanical sorting stage. This either removes recyclable elements from a mixed waste stream (such as metals, plastics, glass and paper) or processes them. It typically involves factory style conveyors, industrial magnets, eddy current separators, trommels, shredders and other tailor made systems, or the sorting is done manually at hand picking stations. The mechanical element has a number of similarities to a materials recovery facility (MRF).

Wet material recovery facility, Hiriya, Israel

Some systems integrate a wet MRF to separate by density and floatation and to recover & wash the recyclable elements of the waste in a form that can be sent for recycling. MBT can alternatively process the waste to produce a high calorific fuel termed refuse derived fuel (RDF). RDF can be used in cement kilns or thermal combustion power plants and is generally made up from plastics and biodegradable organic waste. Systems which are configured to produce RDF include the Herhof and Ecodeco Processes. It is a common misconception that all MBT processes produce RDF. This is not the case and depends strictly on system configuration and suitable local markets for MBT outputs.

Biological Processing

Twin stage & UASB anaerobic digesters

The "biological" element refers to either:

- Anaerobic digestion
- Composting
- Biodrying

Anaerobic digestion harnesses anaerobic microorganisms to break down the biodegradable component of the waste to produce biogas and soil improver. The biogas can be used to generate electricity and heat.

Biological can also refer to a composting stage. Here the organic component is broken down by naturally occurring aerobic microorganisms. They breakdown the waste into carbon dioxide and compost. There is no green energy produced by systems employing only composting treatment for the biodegradable waste.

In the case of biodrying, the waste material undergoes a period of rapid heating through the action of aerobic microbes. During this partial composting stage the heat generated by the microbes result in rapid drying of the waste. These systems are often configured to produce a refuse-derived fuel where a dry, light material is advantageous for later transport and combustion.

Some systems incorporate both anaerobic digestion and composting. This may either take the form of a full anaerobic digestion phase, followed by the maturation (composting) of the digestate. Alternatively a partial anaerobic digestion phase can be induced on water that is percolated through the raw waste, dissolving the readily available sugars, with the remaining material being sent to a windrow composting facility.

By processing the biodegradable waste either by anaerobic digestion or by composting MBT technologies help to reduce the contribution of greenhouse gases to global warming.

Usable wastes for this system:

- Municipal solid waste

- Commercial and industrial waste

- Sewage sludge

Possible products of this system:

- Renewable fuel (biogas) leading to renewable power

- Recovered recyclable materials such as metals, paper, plastics, glass etc.

- Digestate - an organic fertiliser and soil improver

- Carbon credits – additional revenues

- High calorific fraction refuse derived fuel - Renewable fuel content dependent upon biological component

- Residual unusable materials prepared for their final safe treatment (e.g. incineration or gasification) and/or landfill

Further advantages:

- Small fraction of inert residual waste

- Reduction of the waste volume to be deposited to at least a half (density > 1.3 t/m^3), thus the lifetime of the landfill is at least twice as long as usually

- Utilisation of the leachate in the process

- Landfill gas not problematic as biological component of waste has been stabilised

- Daily covering of landfill not necessary

Consideration of Applications

MBT systems can form an integral part of a region's waste treatment infrastructure. These systems are typically integrated with kerbside collection schemes. In the event that a refuse-derived fuel is produced as a by-product then a combustion facility would be required. This could either be an incineration facility or a gasifier.

Alternatively MBT solutions can diminish the need for home separation and kerbside collection of recyclable elements of waste. This gives the ability of local authorities, municipalities and councils to reduce the use of waste vehicles on the roads and keep recycling rates high.

Position of Environmental Groups

Friends of the Earth suggests that the best environmental route for residual waste is to firstly maximise removal of remaining recyclable materials from the waste stream (such as metals, plastics and paper). The amount of waste remaining should be composted or anaerobically digested and disposed of to landfill, unless sufficiently clean to be used as compost.

A report by Eunomia undertook a detailed analysis of the climate impacts of different residual waste technologies. It found that an MBT process that extracts both the metals and plastics prior to landfilling is one of the best options for dealing with our residual waste, and has a lower impact than either MBT processes producing RDF for incineration or incineration of waste without MBT.

Friends of the Earth does not support MBT plants that produce refuse derived fuel (RDF), and believes MBT processes should occur in small, localised treatment plants.

Garbage Truck

Garbage truck or dustcart refers to a truck specially designed to collect municipal solid waste and haul the collected waste to a solid waste treatment facility such as a landfill. Other common names for this type of truck include trash truck in the United States, and rubbish truck, bin wagon, dustbin lorry, bin lorry or bin van elsewhere. Technical names include waste collection vehicle and refuse collection vehicle. These trucks are a common sight in most urban areas.

A Scania front loader

History

Wagons and other means had been used for centuries to haul away solid waste. Among the first self-propelled garbage trucks were those ordered by Chiswick District Council from the Thornycroft Steam Wagon and Carriage Company in 1897 described as a steam motor tip-car, a new design of body specific for "the collection of dust and house refuse".

THORNYCROFT'S STEAM DUST-CART.

Thornycroft Steam Dust-Cart of 1897 with tipper body

The 1920s saw the first open-topped trucks being used, but due to foul odors and waste falling from the back, covered vehicles soon became more common. These covered trucks were first introduced in more densely populated Europe and then in North America, but were soon used worldwide.

The main difficulty was that the waste collectors needed to lift the waste to shoulder height. The first technique developed in the late 20s to solve this problem was to build round compartments with massive corkscrews that would lift the load and bring it away from the rear. A more efficient model was the development of the hopper in 1929. It used a cable system that could pull waste into the truck.

In 1937, George Dempster invented the *Dempster-Dumpster* system in which wheeled waste containers were mechanically tipped into the truck. His containers were known as Dumpsters, which led to the word dumpster entering the language.

In 1938, the Garwood Load Packer revolutionized the industry when the notion of including a compactor in the truck was implemented. The first primitive compactor could double a truck's capacity. This was made possible by use of a hydraulic press which compacted the contents of the truck periodically.

RS-3 Lightning Rear Steer truck

1955 saw the Dempster Dumpmaster the first front loader introduced, however they didn't become common until the 1970s. The 1970s also saw the introduction of smaller dumpsters, often known as wheelie bins which were also emptied mechanically. Since that time there has been little dramatic change, although there have been various improvements to the compaction mechanisms in order to improve payload. In the mid-1970s Petersen Industries introduced the first grapple truck for municipal waste collection.

In 1997, Lee Rathbun introduced the *Lightning Rear Steer System*. This system includes an elevated, rear-facing cab for both driving the truck and operating the loader. This configuration allows the operator to follow behind haul trucks and load continuously.

Types of Waste Collection Vehicle

A standard Waste Management Inc. front-loading garbage truck in San Jose, California

Front Loaders

Front loaders generally service commercial and industrial businesses using large waste containers with lids known as Dumpsters in the US. The truck is equipped with powered forks on the front which the driver carefully aligns with sleeves on the waste container using a joystick or a set of levers. The waste container is then lifted over the truck. Once it gets to the top the container is then flipped upside down and the waste or recyclable material is emptied into the vehicle's hopper. Once the waste is dumped, it is compacted by a hydraulically powered moving wall that oscillates backwards and forwards to push the waste to the rear of the vehicle. Most of the newer packing trucks have "pack-on-the-go hydraulics" which lets the driver pack loads while driving, allowing faster route times. When the body is full, the compaction wall moves all the way to the rear of the body, ejecting it via an open tailgate. There is also a system called the Curotto Can which is an attachment for a front loader that has an automated arm that functions as an automated side loader that allows the driver to dump carts.

14.5 m³ rear load container serviced in Copenhagen

Garbagemen loading garbage by hand in Japan, 2013

Rear loaders

Rear loaders have an opening at the rear that a waste collector can throw waste bags or empty the contents of bins into. Often in many areas they have a lifting mechanism to automatically empty large carts without the operator having to lift the waste by hand.

Another popular system for the rear loader is a rear load container specially built to fit a groove in the truck. The truck will have a chain or cable system for upending the container. The waste will then slide into the hopper of the truck.

The modern rear loader usually compacts the waste using a hydraulically powered mechanism that employs a moving plate or shovel to scoop the waste out from the loading hopper and compress it against a moving wall. In most compactor designs, the plate has a pointed edge (hence giving it the industry standard name *packer blade*) which is designed to apply point pressure to the waste to break down bulky items in the hopper before being drawn into the main body of the truck.

Compactor designs, however, have been many and varied, however the two most popular in use today are the "sweep and slide" system (first pioneered on the Leach 2R Packmaster), where the packer blade pivots on a moving carriage which slides back and forth, and the "swing link" system (such as the Dempster Routechief) where the blade literally swings on a "pendulum"-style mechanism. The Heil Colectomatic used a combination of a lifting loading hopper and a pivoting sweeper blade to clear and compact waste in anticipation of the next load.

So-called "continuous" compactors were popular in the 1960s and 1970s. The German Shark design (later Rotopress) used a huge rotating drum, analogous to a cement mixer, in conjunction with a serrated auger to grind down and compact the garbage. SEMAT-Rey of France pioneered the rotating rake system (also used in the British Shelvoke and Drewry Revopak) to both mutilate waste and break down large items. Other systems used a large Archimedes' screw to draw in waste and mutilate it inside the body. A mixture of safety concerns, and higher fuel consumption has seen a decline in the popularity of continuously compacting garbage trucks. The Rotopress design remains popular due to its niche in being able to effectively deal with green waste for composting.

The wall will move towards the front of the vehicle as the pressure forces the hydraulic valves to open, or as the operator moves it with a manual control.

A unique rear-loading system involves a rear loader and a front-loading tractor (usually a Caterpillar front loader with a Tink Claw) for yard waste collection (and in some cities, garbage and recycling). The front loader picks up yard waste set in the street, and then loaded into the back of a rear loader. This system is used in several cities, including San Jose.

Side Loaders

Side loaders are loaded from the side, either manually, or with the assistance of a joystick-controlled robotic arm with a claw, used to automatically lift and tip wheeled bins into the truck's hopper.

Automated Side Loaders

Lift-equipped trucks are referred to as automated side loaders, or ASL's. Similar to a front-end loader, the waste is compacted by an oscillating packer plate at the front of the loading hopper which forces the waste through an aperture into the main body and is therefore compacted towards the rear of the truck. An Automated Side Loader only needs one operator, where a traditional rear load garbage truck may require two or three people, and has the additional advantage of reducing on the job injuries due to repetitive heavy lifting. Due to these advantages, ASL's have become more popular than traditional manual collection. Typically an Automated Side Loader uses standardized wheeled carts compatible with the truck's automated lift.

Semi-automated Side Loaders

Semi-automated side loaders use an automated mechanism to lift and dump manually aligned waste containers inside the main body of the truck. The primary difference of semi-automated side loaders is that they require more than one person, to operate the truck, and to manually bring and align containers to the loading hopper on the side of the truck.

An Automated side loader garbage truck in Canberra, Australia

As with front loaders, the compaction mechanism comprises a metal pusher plate in the collection hopper which oscillates backwards and forwards under hydraulic pressure, pushing the refuse through an aperture, thus compacting it against the material already loaded. On some ASL's there is also a "folding" crusher plate positioned above the opening in the hopper, that folds down to crush bulky items within reach of the metal pusher plate. Another compactor design is the "paddle packer" which uses a paddle that rotates from side to side, forcing refuse into the body of the truck.

Automated garbage collection in Aardenburg, Netherlands

Automated side loader operation

Garbage collection by an automatic side loader during autumn in Kelowna, British Columbia, Canada

Pneumatic Collection

Volvo pneumatic collector used for "waste suction"

Pneumatic collection trucks have a crane with a tube and a mouthpiece that fits in a hole, usually hidden under a plate under the street. From here it will suck up waste from an underground installation. The system usually allows the driver to "pick up" the waste, even if the access is blocked by cars, snow or other barriers.

Grapple truck

Grapple Trucks

Grapple trucks enable the collection of bulk waste. A large percentage of items in the solid waste stream are too large or too heavy to be safely lifted by hand into traditional garbage trucks. These items (furniture, large appliances, branches, logs) are called bulky waste or "oversized." The preferred method for collecting these items is with a grapple truck. Grapple trucks have hydraulic knucklebooms, tipped with a clamshell bucket, and usually include a dump body or trailer.

Roll-offs

Roll-offs are characterized by a rectangular footprint, utilizing wheels to facilitate rolling the dumpster in place. The container is designed to be transported by special roll-off trucks. They are relatively efficient for bulk loads of waste.

Sand Cleaning Machine

A sand cleaning machine, beach cleaner, or (colloquially) sandboni is a vehicle that drags a raking or sifting device over beach sand to remove rubbish and other foreign matter. They are manually self-pulled vehicles on tracks or wheels or pulled by quad-bike or tractor. Seaside cities use beach cleaning machines to combat the problems of litter left by beach patrons and other pollution washed up on their shores. A chief task in beach cleaning strategies is finding the best way to handle waste matter on the beaches, taking into consideration beach erosion and changing terrain. Beach cleaning machines work by collecting sand by way of a scoop or drag mechanism and then raking or sifting anything large enough to be considered foreign matter, including sticks, stones, rubbish, syringes and other items. Similar applications include lake beaches, sandfields for beach volleyball and kindergarten and playing field sandpits. The word "sandboni" is a back-formation referencing the ice-surfacing machine Zamboni.

Raking beach cleaner purifies sand

A tractor-pulled beach cleaner at Hietaniemi beach in Helsinki, Finland.

Common Technologies

Raking technology can be used on dry or wet sand. When using this method, a rotating conveyor belt containing hundreds of tines combs through the sand and removes surface and buried debris while leaving the sand on the beach. Raking machines can remove materials ranging in size from small pebbles, shards of glass, and cigarette butts to larger debris, like seaweed and driftwood. By keeping the sand on the beach and only lifting the debris, raking machines can travel at high speeds.

Sifting technology is practiced on dry sand and soft surfaces. The sand and waste are collected via the pick-up blade of the vehicle onto a vibrating screening belt, which leaves the sand behind. The waste is gathered in a collecting tray which is often situated at the back of the vehicle. Because sand and waste are lifted onto the screening belt, sifters must allow time for the sand to sift through the screen and back onto the beach. The size of the materials removed is governed by the size of the holes in the installed screen.

Combined raking and sifting technology differs from pure sifters in that it uses rotating tines to scoop sand and debris onto a vibrating screen instead of relying simply on the pick-up blade. The tines' position can be adjusted to more effectively guide different-sized materials onto the screen. Once on the screen, combined raking and sifting machines use the same technology as normal sifters to remove unwanted debris from the sand.

Sand sifting by hand is used for smaller areas or sensitive habitat. Sand and debris is collected into a windrow or pile and manually shoveled onto screened sifting trays to separate the debris from the sand. While effective, it requires the movement of sand to the site of the tray, and then redistribution of the sand after sifting. A more efficient method is the use of a screened fork at the place where the debris is located. The effort to manually agitate the sand can become tiresome; however, a recent development of a battery-powered sand rake combines the spot cleaning effectiveness of manual screening with the ease of an auto-sifting hand tool.

Areas of Operation

Sand cleaning machines are used all over the world to ensure the safety and happiness of beach-goers. By removing litter, unwanted seaweed, and other debris from the beach, municipalities and resorts are able to maintain their beaches with fewer invested hours.

In addition to their regular litter-removing uses, beach and sand cleaners have been used to clean up after natural disasters. For example:

In Galveston, Texas, low oxygen levels in the water resulted in thousands of dead fish washing ashore. Raking sand cleaners were then used to remove the rotting fish off the beach before they released excessive toxins into the air, sand, and water.

Sandboni at the gulf of Mexico

The Olympic Games 2008 saw the first remote-control Sandbonis for the beach volley-ball fields in Beijing Chaoyang Park.

Beach volleyball field after rework of a sand cleaner

The cleanup after the Deepwater Horizon oil spill saw large applications of sand cleaners to the area. Similarly, the Rena oil spill in New Zealand also saw beach cleaners deployed in an effort to remove the affected sand.

Manufacturers

The major manufacturers of large beach-sand cleaning machines are considered to be H Barber & Sons, Cherrington, Beach Tech, Rockland and Tirrenia Srl.

There are many other manufacturers of sand cleaners being used for other purposes. For example, a smaller 4-wheel and halftrack sand cleaning machine is used for sandpits in Kindergarten and municipality playing fields and for beach volleyball. When environmental or spot-cleaning requires hand operations, an auto-sifting, lightweight, screened rake can be the best choice

Waste-to-energy

Waste-to-energy (WtE) or energy-from-waste (EfW) is the process of generating energy in the form of electricity and/or heat from the primary treatment of waste. WtE is a form of energy recovery. Most WtE processes produce electricity and/or heat directly through combustion, or produce a combustible fuel commodity, such as methane, methanol, ethanol or synthetic fuels.

Spittelau incineration plant is one of several plants that provide district heating in Vienna.

History

The first incinerator or "Destructor" was built in Nottingham UK in 1874 by Manlove, Alliott & Co. Ltd. to the design of Albert Fryer.

The first US incinerator was built in 1885 on Governors Island in New York, NY.

The first waste incinerator in Denmark was built in 1903 in Frederiksberg

The first facility in Czech Republic was built in 1905 in Brno.

Incineration

Incineration, the combustion of organic material such as waste with energy recovery, is the most common WtE implementation. All new WtE plants in OECD countries incinerating waste (residual MSW, commercial, industrial or RDF) must meet strict emission standards, including those on nitrogen oxides (NO_x), sulphur dioxide (SO_2), heavy metals and dioxins. Hence, modern incineration plants are vastly different from old types, some of which neither recovered energy nor materials. Modern incinerators reduce the volume of the original waste by 95-96 percent, depending upon composition and degree of recovery of materials such as metals from the ash for recycling.

Incinerators may emit fine particulate, heavy metals, trace dioxin and acid gas, even though these emissions are relatively low from modern incinerators. Other concerns include proper management of residues: toxic fly ash, which must be handled in hazardous waste disposal installation as well as incinerator bottom ash (IBA), which must be reused properly.

Critics argue that incinerators destroy valuable resources and they may reduce incentives for recycling. The question, however, is an open one, as countries in Europe recycling the most (up to 70%) also incinerate their residual waste to avoid landfilling.

Incinerators have electric efficiencies of 14-28%. In order to avoid losing the rest of the energy, it can be used for e.g. district heating (cogeneration). The total efficiencies of cogeneration incinerators are typically higher than 80% (based on the lower heating value of the waste).

The method of using incineration to convert municipal solid waste (MSW) to energy is a relatively old method of WtE production. Incineration generally entails burning waste (residual MSW, commercial, industrial and RDF) to boil water which powers steam generators that make electric energy and heat to be used in homes, businesses, institutions and industries. One problem associated with incinerating MSW to make electrical energy is the potential for pollutants to enter the atmosphere with the flue gases from the boiler. These pollutants can be acidic and in the 1980s were reported to cause environmental damage by turning rain into acid rain. Since then, the industry has removed this problem by the use of lime scrubbers and electro-static precipitators on smokestacks. By passing the smoke through the basic lime scrubbers, any acids that might be in the smoke are neutralized which prevents the acid from reaching the atmosphere and hurting the environment. Many other devices, such as fabric filters, reactors, and catalysts destroy or capture other regulated pollutants. According to the New York Times, modern incineration plants are so clean that "many times more dioxin is now released from home fireplaces and backyard barbecues than from incineration. " According to the German Environmental Ministry, "because of stringent regulations, waste incineration plants are no longer significant in terms of emissions of dioxins, dust, and heavy metals".

WTE Technologies Other than Incineration

There are a number of other new and emerging technologies that are able to produce energy from waste and other fuels without direct combustion. Many of these technologies have the potential to produce more electric power from the same amount of fuel than would be possible by direct combustion. This is mainly due to the separation of corrosive components (ash) from the converted fuel, thereby allowing higher combustion temperatures in e.g. boilers, gas turbines, internal combustion engines, fuel cells. Some are able to efficiently convert the energy into liquid or gaseous fuels:

Thermal technologies:

- Gasification: produces combustible gas, hydrogen, synthetic fuels

- Thermal depolymerization: produces synthetic crude oil, which can be further refined

- Pyrolysis: produces combustible tar/biooil and chars

- Plasma arc gasification or plasma gasification process (PGP): produces rich syngas including hydrogen and carbon monoxide usable for fuel cells or generating electricity to drive the plasma arch, usable vitrified silicate and metal ingots, salt and sulphur

Non-thermal technologies:

- Anaerobic digestion: Biogas rich in methane

- Fermentation production: examples are ethanol, lactic acid, hydrogen

- Mechanical biological treatment (MBT)

  o MBT + Anaerobic digestion

  o MBT to Refuse derived fuel

Global WTE Developments

During the 2001-2007 period, the WTE capacity increased by about four million metric tons per annum. Japan and China built several plants that were based on direct smelting or on fluidized bed combustion of solid waste. In China there are about 50 WTE plants. Japan is the largest user in thermal treatment of MSW in the world with 40 million tons. Some of the newest plants use stoker technology and others use the advanced oxygen enrichment technology. There are also over one hundred thermal treatment plants using relatively novel processes such as direct smelting, the Ebara fluidization process and the Thermo- select -JFE gasification and melting technology process. In Patras, Greece, a Greek company just finished testing a system that shows potential. It generates 25kwatts of electricity and 25kwatts of heat from waste water. In India its first energy bio-science center was developed to reduce the country's green house gases and its dependency on fossil fuel. As of June 2014, Indonesia had a total of 93.5MW installed capacity of WtE, with a pipeline of projects in different preparation phases together amounting to another 373MW of capacity.

Biofuel Energy Corporation of Denver, CO, opened two new biofuel plants in Wood River, NE, and Fairmont, MN, in July 2008. These plants use distillation to make ethanol for use in motor vehicles and other engines. Both plants are currently reported to be working at over 90% capacity. Fulcrum BioEnergy incorporated located in Pleasanton, CA, is currently building a WTE plant near Reno, NV. The plant is scheduled to open in early 2010 under the name of Sierra BioFuels plant. BioEnergy incorporated predicts that the plant will produce approximately 10.5 million gallons per year of ethanol from nearly 90,000 tons per year of MSW.(Biofuels News)

Waste to energy technology includes fermentation, which can take biomass and create ethanol, using waste cellulosic or organic material. In the fermentation process, the sugar in the waste is changed to carbon dioxide and alcohol, in the same general process that is used to make wine. Normally fermentation occurs with no air present. Esterification can also be done using waste to energy technologies, and the result of this process is biodiesel. The cost effectiveness of esterification will depend on the feedstock being used, and all the other relevant factors such as transportation distance, amount of oil present in the feedstock, and others. Gasification and pyrolysis by now can reach gross thermal conversion efficiencies (fuel to gas) up to 75%, however a complete combustion is superior in terms of fuel conversion efficiency. Some pyrolysis processes need an outside heat source which may be supplied by the gasification process, making the combined process self-sustaining.

Carbon Dioxide Emissions

In thermal WtE technologies, nearly all of the carbon content in the waste is emitted as carbon dioxide (CO_2) to the atmosphere (when including final combustion of the products from pyrolysis and gasification; except when producing bio-char for fertilizer). Municipal solid waste (MSW) contain approximately the same mass fraction of carbon as CO_2 itself (27%), so treatment of 1 metric ton (1.1 short tons) of MSW produce approximately 1 metric ton (1.1 short tons) of CO_2.

In the event that the waste was landfilled, 1 metric ton (1.1 short tons) of MSW would produce approximately 62 cubic metres (2,200 cu ft) methane via the anaerobic decomposition of the biodegradable part of the waste. This amount of methane has more than twice the global warming potential than the 1 metric ton (1.1 short tons) of CO_2, which would have been produced by combustion. In some countries, large amounts of landfill gas are collected, but still the global warming potential of the landfill gas emitted to atmosphere in e.g. the US in 1999 was approximately 32% higher than the amount of CO_2 that would have been emitted by combustion.

In addition, nearly all biodegradable waste is biomass. That is, it has biological origin. This material has been formed by plants using atmospheric CO_2 typically within the last growing season. If these plants are regrown the CO_2 emitted from their combustion will be taken out from the atmosphere once more.

Such considerations are the main reason why several countries administrate WtE of the biomass part of waste as renewable energy. The rest—mainly plastics and other oil and gas derived products—is generally treated as non-renewables.

Determination of the Biomass Fraction

MSW to a large extent is of biological origin (biogenic), e.g. paper, cardboard, wood, cloth, food scraps. Typically half of the energy content in MSW is from biogenic material. Consequently, this energy is often recognised as renewable energy according to the waste input.

Several methods have been developed by the European CEN 343 working group to determine the biomass fraction of waste fuels, such as Refuse Derived Fuel/Solid Recovered Fuel. The initial two methods developed (CEN/TS 15440) were the manual sorting method and the selective dissolution method. A detailed systematic comparison of these two methods was published in 2010. Since each method suffered from limitations in properly characterizing the biomass fraction, two alternative methods have been developed.

The first method uses the principles of radiocarbon dating. A technical review (CEN/TR 15591:2007) outlining the carbon 14 method was published in 2007. A technical standard of the carbon dating method (CEN/TS 15747:2008) will be published in 2008. In the United States, there is already an equivalent carbon 14 method under the standard method ASTM D6866.

The second method (so-called balance method) employs existing data on materials composition and operating conditions of the WtE plant and calculates the most probable result based on a mathematical-statistical model. Currently the balance method is installed at three Austrian and eight Danish incinerators.

A comparison between both methods carried out at three full-scale incinerators in Switzerland showed that both methods came to the same results.

Carbon 14 dating can determine with precision the biomass fraction of waste, and also determine the biomass calorific value. Determining the calorific value is important for green certificate programs such as the Renewable Obligation Certificate program in the United Kingdom. These programs award certificates based on the energy produced from biomass. Several research papers, including the one commissioned by the Renewable Energy Association in the UK, have been published that demonstrate how the carbon 14 result can be used to calculate the biomass calorific value. The UK gas and electricity markets authority, Ofgem, released a statement in 2011 accepting the use of Carbon 14 as a way to determine the biomass energy content of waste feedstock under their administration of the Renewables Obligation. Their Fuel Measurement and Sampling (FMS) questionnaire describes the information they look for when considering such proposals.

Examples of Waste-to-energy Plants

According to ISWA there are 431 WtE plants in Europe (2005) and 89 in the United States (2004). The following are some examples of WtE plants.

Waste incineration WtE plants

- Essex County Resource Recovery Facility, Newark, New Jersey

- Lee County Solid Waste Resource Recovery Facility, Fort Myers, Florida, USA (1994)

- Montgomery County Resource Recovery Facility in Dickerson, Maryland, USA (1995)

- Spittelau (1971), and Flötzersteig (1963), Vienna, Austria (Wien Energie)

- SYSAV waste-to-energy plant in Malmö (2003 and 2008), Sweden (Flash presentation)

- Algonquin Power, Brampton, Ontario, Canada

- Stoke Incinerator, Stoke-on-Trent, UK (1989)

- Teesside EfW plant near Middlesbrough, North East England (1998)

- Edmonton Incinerator in Greater London, England (1974)

- Burnaby Waste-to-Energy Facility, Metro Vancouver, Canada (1988)

- Timarpur-Okhla Waste to Energy Plant, New Delhi, India

Liquid fuel producing plants (planned or under construction)

- Edmonton Waste-to-ethanol Facility located in Edmonton, Alberta, Canada based on the Enerkem-process, fueled by RDF. Initially scheduled for completion during 2010 commissioning of front-end systems commenced December 2013 and Enerkem then expected initial methanol production during 2014. Production start has been delayed several times. As of spring 2016 Enerkem expected ethanol production to commence some time in 2017, and no public confirmation of any actual RDF processing was available.

- Mississippi Waste-to-ethanol Plant, Enerkem-process, initially scheduled for completion 2013. Pontotoc, Mississippi, USA. As of February 2016, the Mississippi project had silently disappeared from Enerkem's list of projects.

Plasma Gasification Waste-to-Energy plants

- The US Air Force once tested a Transportable Plasma Waste to Energy System (TPWES) facility (PyroGenesis technology) at Hurlburt Field, Florida. The plant, which cost cost $7.4 million to construct, was closed and sold at a government liquidation auction in May 2013, less than three years after its commissioning. The opening bid was $25. The winning bid was sealed.

Besides large plants, domestic waste-to-energy incinerators also exist. For example, the refuge de Sarenne has a domestic waste-to-energy plant. It is made by combining a wood-fired gasification boiler with a Stirling motor.

Waste Collection

Waste collection is a part of the process of waste management. It is the transfer of solid waste from the point of use and disposal to the point of treatment or landfill. Waste collection also includes the curbside collection of recyclable materials that technically are not waste, as part of a municipal landfill diversion program.

A waste collection vehicle in Sakon Nakhon, Thailand.

Manual waste collection in Bukit Batok West, Singapore.

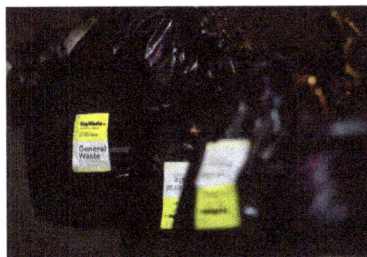

Waste on a sidewalk for collection, bagged and stickered - in Dublin, Ireland

Household Waste

Household waste in economically developed countries will generally be left in waste containers or recycling bins prior to collection by a waste collector using a waste collection vehicle.

However, in many developing countries, such as Mexico and Egypt, waste left in bins or bags at the side of the road will not be removed unless residents interact with the waste collectors.

Mexico City residents must haul their trash to a waste collection vehicle which makes frequent stops around each neighborhood. The waste collectors will indicate their readiness by ringing a distinctive bell and possibly shouting. Residents line up and hand their trash container to the waste collector. A tip may be expected in some neighborhoods. Private contracted waste collectors may circulate in the same neighborhoods as many as five times per day, pushing a cart with a waste container, ringing a bell and shouting to announce their presence. These private contractors are not paid a salary, and survive only on the tips they receive. Later, they meet up with a waste collection vehicle to deposit their accumulated waste.

The waste collection vehicle will often take the waste to a transfer station where it will be loaded up into a larger vehicle and sent to either a landfill or alternative waste treatment facility.

Commercial Waste

Waste collection considerations include type and size of bins, positioning of the bins, and how often bins are to be serviced. Overfilled bins result in rubbish falling out while being tipped. Hazardous rubbish like empty petrol cans can cause fires igniting other trash when the truck compactor is operating. Bins may be locked or stored in secure areas to avoid having non-paying parties placing rubbish in the bin.

Employment as Garbage Collector

According to USNews

> The Bureau of Labor Statistics predicts that employment in this industry will increase 16.2 percent, adding 21,600 new jobs, by 2022. The main drivers of this growth include a rise in population, individual income and more people choosing to recycle.

Waste Sorting

Waste sorting is the process by which waste is separated into different elements. Waste sorting can occur manually at the household and collected through curbside collection schemes, or automatically separated in materials recovery facilities or mechanical biological treatment systems. Hand sorting was the first method used in the history of waste sorting.

Recycling bins in Singapore

Manual waste sorting for recycling

Garbage containers in Fuchū, Tokyo, Japan

Waste can also be sorted in a civic amenity site.

Waste segregation means dividing waste into dry and wet. Dry waste includes wood and related products, metals and glass. Wet waste, typically refers to organic waste usually generated by eating establishments and are heavy in weight due to dampness. Waste can also be segregated on basis of biodegradable or non-biodegradable waste.

Characteristic containers for recycling in Portovenere, Italy

Landfills are an increasingly pressing problem. Less and less land is available to deposit refuse, but the volume of waste is growing all time. As a result, segregating waste is not just of environmental importance, but of economic concern, too.

Methods

Waste is collected at its source in each area and separated. The way that waste is sorted must reflect local disposal systems. The following categories are common:

- Paper

- Cardboard (including packaging for return to suppliers)

- Glass (clear, tinted – no light bulbs or window panes, which belong with residual waste)

- Plastics

- Textiles

- Wood, leather, rubber

- Scrap metal

- Compost

- Special/hazardous waste

- Residual waste

Organic waste can also be segregated for disposal:

- Leftover food which has had any contact with meat can be collected separately to prevent the spread of bacteria.

  o Meat and bone can be retrieved by bodies responsible for animal waste

   o   If other leftovers are sent, for example, to local farmers, they can be ster-
       ilised before being fed to the animals

- Peel and scrapings from fruit and vegetables can be composted along with other
  degradable matter. Other waste can be included for composting, too, such as cut
  flowers, corks, coffee grindings, rotting fruit, tea bags, egg- and nutshells, paper
  towels etc.

Chip pan oil (fryer oil), used fats, vegetable oil and the content of fat filters can be collected by companies able to re-use them. Local authority waste departments can provide relevant addresses. This can be achieved by providing recycling bins.

By Country

In Germany, regulations exist that provide mandatory quotas for the waste sorting of
packaging waste and recyclable materials such as glass bottles.

In Denpasar, Bali, Indonesia, a pilot project using an automated collecting machine
of plastic bottles or aluminium cans with voucher reward has been implemented in a
market.

References

- Ministry for the Environment (December 2007). Environment New Zealand 2007. Ministry for
  the Environment (New Zealand). ISBN 978-0-478-30192-2. Retrieved 2008-03-27.

- "Waste Pesticide Management" (PDF). Oregon.Gov. State of Oregon Department of Environmental Quality Land Quality Division Hazardous Waste Program. Retrieved 5 October 2016.

- "Cost Effective Waste to Energy Technologies – Updated Article With Extra Information". bion-
  omicfuel.com. Retrieved 28 February 2015.

- "Autonomie énergétique pour un refuge de montagne : panneaux solaires". Connaissance des
  Énergies. 5 July 2012. Retrieved 28 February 2015.

- Frommer, Dan. "Cities as Gadgets: 8 Features This Brand-New City Has That Yours Doesn't".
  Readwrite. Retrieved 3 August 2015.

- Normandin, Pierre-Andre. "Projet de collecte des déchets: trois millions aux poubelles". La Pres-
  se. Retrieved 3 August 2015.

- Clarke, Katherine (June 27, 2014). "TOO RICH FOR TRASH: Hudson Yards waste will exit by
  pneumatic tube". NY Daily News. Retrieved June 27, 2014.

- Geroux, Zachary; Voytko, Eric. "The Ever Expanding History of the Front Load Refuse Truck".
  Retrieved 10 September 2014.

- "AFSOC makes 'green' history while investing in future". US Air Force Special Operations Command. Retrieved 2011-04-28. .

Different Disposal Practices

This chapter focuses on different practices of disposing waste. Some of these practices are landfill, incineration and compost. Waste materials have certain organic substances. The burning of these organic substances is known as incineration. Landfills on the other hand are locations, which are used for the removal of waste materials by burial.

Landfill

A landfill site (also known as a tip, dump, rubbish dump, garbage dump or dumping ground and historically as a midden) is a site for the disposal of waste materials by burial and is the oldest form of waste treatment (although the burial part is modern; historically, refuse was just left in piles or thrown into pits). Historically, landfills have been the most common method of organized waste disposal and remain so in many places around the world.

A landfill in Poland

Some landfills are also used for waste management purposes, such as the temporary storage, consolidation and transfer, or processing of waste material (sorting, treatment, or recycling).

A landfill also may refer to ground that has been filled in with rocks instead of waste materials, so that it can be used for a specific purpose, such as for building houses. Unless they are stabilized, these areas may experience severe shaking or soil liquefaction of the ground during a large earthquake.

Operations

Typically, operators of well-run landfills for non-hazardous waste meet predefined specifications by applying techniques to:

One of several landfills used by Dryden, Ontario, Canada.

1. confine waste to as small an area as possible

2. compact waste to reduce volume

3. cover waste (usually daily) with layers of soil

During landfill operations a scale or weighbridge may weigh waste-collection vehicles on arrival and personnel may inspect loads for wastes that do not accord with the landfill's waste-acceptance criteria. Afterward, the waste-collection vehicles use the existing road network on their way to the tipping face or working front, where they unload their contents. After loads are deposited, compactors or bulldozers can spread and compact the waste on the working face. Before leaving the landfill boundaries, the waste collection vehicles may pass through a wheel-cleaning facility. If necessary, they return to the weighbridge for re-weighing without their load. The weighing process can assemble statistics on the daily incoming waste-tonnage, which databases can retain for record keeping. In addition to trucks, some landfills may have equipment to handle railroad containers. The use of "rail-haul" permits landfills to be located at more remote sites, without the problems associated with many truck trips.

Typically, in the working face, the compacted waste is covered with soil or alternative materials daily. Alternative waste-cover materials include chipped wood or other "green waste", several sprayed-on foam products, chemically "fixed" bio-solids, and temporary blankets. Blankets can be lifted into place at night and then removed the following day prior to waste placement. The space that is occupied daily by the compacted waste and the cover material is called a daily cell. Waste compaction is critical to extending the life of the landfill. Factors such as waste compressibility, waste-layer thickness and the number of passes of the compactor over the waste affect the waste densities.

Advantages

Landfills are often the most cost-efficient way to dispose of waste, especially in countries like the United States with large open spaces. While resource recovery and incineration both require extensive investments in infrastructure, and material recovery also requires extensive manpower to maintain, landfills have fewer fixed—or ongoing—costs, allowing them to compete favorably. In addition, landfill gas can be upgraded to natural gas—landfill gas utilization—which is a potential revenue stream.

Social and Environmental Impact

Landfills have the potential to cause a number of issues. Infrastructure disruption, such as damage to access roads by heavy vehicles, may occur. Pollution of local roads and water courses from wheels on vehicles when they leave the landfill can be significant and can be mitigated by wheel washing systems. Pollution of the local environment, such as contamination of groundwater or aquifers or soil contamination may occur, as well.

Landfill operation in Hawaii. Note that the area being filled is a single, well-defined "cell" and that a protective landfill liner is in place (exposed on the left) to prevent contamination by leachates migrating downward through the underlying geological formation.

Leachate

Extensive efforts are made to capture and treat leachate from landfills before it reaches groundwater aquifers, but engineered liners always have a lifespan, though it may be 100 years or more. Eventually, every landfill liner will leak, allowing the leachate to contaminate the groundwater. Installation of composite liners with flexible membrane and soil barrier is enforced by the EPA to ensure that leachate is withheld.

Dangerous Gases

Methane is naturally generated by decaying organic wastes in a landfill. It is a potent greenhouse gas, and can itself be a danger because it is flammable and potentially explosive. In properly managed landfills, gas is collected and utilized. This could range from simple flaring to landfill gas utilization.

Infections

Poorly run landfills may become nuisances because of vectors such as rats and flies which can cause infectious diseases. The occurrence of such vectors can be mitigated through the use of daily cover.

Other potential issues include wildlife disruption, dust, odor, noise pollution, and reduced local property values.

Landfill Gas

Gases are produced in landfills due to the anaerobic digestion by microbes. In a properly managed landfill this gas is collected and used. Its uses range from simple flaring to the landfill gas utilization and generation of electricity. Landfill gas monitoring alerts workers to the presence of a build-up of gases to a harmful level. In some countries, landfill gas recovery is extensive; in the United States, for example, more than 850 landfills have active landfill gas recovery systems.

Regional Practice

A landfill in Perth, Western Australia

South East New Territories Landfill, Hong Kong

Canada

Landfills in Canada are regulated by provincial environmental agencies and environmental protection acts (EPA). Older facilities tend to fall under current standards and are monitored for leaching. Some former locations have been converted to parkland.

European Union

In the European Union, individual states are obliged to enact legislation to comply with the requirements and obligations of the European Landfill Directive. In the UK this is the Waste Implementation Programme.

United Kingdom

Landfilling practices in the UK have had to change in recent years to meet the challenges of the European Landfill Directive. The UK now imposes landfill tax upon biodegradable waste which is put into landfills. In addition to this the Landfill Allowance Trading Scheme has been established for local authorities to trade landfill quotas in England. A different system operates in Wales where authorities are not able to 'trade' between themselves, but have allowances known as the Landfill Allowance Scheme.

United States

U.S. landfills are regulated by each state's environmental agency, which establishes minimum guidelines; however, none of these standards may fall below those set by the United States Environmental Protection Agency (EPA).

Permitting a landfill generally takes between 5 and 7 years, costs millions of dollars and requires rigorous siting, engineering and environmental studies and demonstrations to ensure local environmental and safety concerns are satisfied.

Microbial Topics

The status of a landfill's microbial community may determine its digestive efficiency.

Bacteria that digest plastic have been found in landfills.

Reclaiming Materials

Landfills can be regarded as a viable and abundant source of materials and energy. In the developing world, waste pickers often scavenge for still-usable materials. In a commercial context, landfill sites have also been discovered by companies, and many have begun harvesting materials and energy . Well known examples are gas recovery facilities. Other commercial facilities include waste incinerators which have built-in material recovery. This material recovery is possible through the use of filters (electro filter, active carbon and potassium filter, quench, HCl-washer, SO_2-washer, bottom ash-grating, etc.).

Alternatives

In addition to waste reduction and recycling strategies, there are various alternatives to landfills, including Waste-to-energy incineration, anaerobic digestion, composting, mechanical biological treatment, pyrolysis and plasma arc gasification, which have all begun to establish themselves in the market. Depending on local economics and incentives, these can be made more financially attractive than landfills.

Restrictions

Countries including Germany, Austria, Sweden, Denmark, Belgium, the Netherlands, and Switzerland, have banned the disposal of untreated waste in landfills. In these countries, only the ashes from incineration or the stabilized output of mechanical biological treatment plants may still be deposited.

Types of Landfill

Bioreactor landfill

Landfills are the primary method of waste disposal in many parts of the world, including United States and Canada. Bioreactor landfills are expected to reduce the amount of and costs associated with management of leachate, to increase the rate of production of methane (natural gas) for commercial purposes and reduce the amount of land required for land-fills. Bioreactor landfills are monitored and manipulate oxygen and moisture levels to increase the rate of decomposition by microbial activity.

Traditional Landfills and Associated Problems

Landfills are the oldest known method of waste disposal. Waste is buried in large dug out pits (unless naturally occurring locations are available) and covered. Bacteria and archaea decompose the waste over several decades producing several by-products of importance, including methane gas (natural gas), leachate, and volatile organic compounds (such as hydrogen sulfide (H_2S), N_2O_2, etc.).

Methane gas, a strong greenhouse gas, can build up inside the landfill leading to an explosion unless released from the cell. Leachate are fluid metabolic products from decomposition and contain various types of toxins and dissolved metallic ions. If leachate escapes into the ground water it can cause health problems in both animals and plants. The volatile organic compounds (VOCs) are associated with causing smog and acid rain. With the increasing amount of waste produced, appropriate places to safely store it have become difficult to find.

Working of a Bioreactor Landfill

There are three types of bioreactor: aerobic, anaerobic and a hybrid (using both aerobic and anaerobic method). All three mechanisms involve the reintroduction of collected leachate supplemented with water to maintain moisture levels in the landfill. The micro-organisms responsible for decomposition are thus stimulated to decompose at an increased rate with an attempt to minimise harmful emissions.

In aerobic bioreactors air is pumped into the landfill using either vertical or horizontal system of pipes. The aerobic environment decomposition is accelerated and amount of VOCs, toxicity of leachate and methane are minimised. In anaerobic bioreactors with leachate being circulated the landfill produces methane at a rate much faster and earlier than traditional landfills. The high concentration and quantity of methane allows it to be used more efficiently for commercial purposes while reducing the time that the landfill needs to be monitored for methane production. Hybrid bioreactors subject the upper portions of the landfill through aerobic-anaerobic cycles to increase decomposition rate while methane is produced by the lower portions of the landfill. Bioreactor landfills produce lower quantities of VOCs than traditional landfills, except H_2S. Bioreactor landfills produce higher quantities of H_2S. The exact biochemical pathway responsible for this increase is not well studied

Advantages of Bioreactor Landfills

Bioreactor landfills accelerate the process of decomposition. As decomposition progresses, the mass of biodegradable components in the landfill declines, creating more space for dumping garbage. Bioreactor landfills are expected to increase this rate of decomposition and save up to 30% of space needed for landfills. With increasing amounts of solid waste produced every year and scarcity of landfill spaces, bioreactor landfill can thus provide a significant way of maximising landfill space. This is not just cost effective, but since less land is needed for the landfills, this is also better for the environment.

Furthermore, most landfills are monitored for at least 3 to 4 decades to ensure that no leachate or landfill gases escape into the community surrounding the landfill site. In contrast, bioreactor landfill are expected to decompose to level that does not require monitoring in less than a decade. Hence, the landfill land can be used for other purposes such as reforestation or parks, depending on the location at an earlier date. In addition, re-using leachate to moisturise the landfill filters it. Thus, less time and energy is required to process the leachate, making the process more efficient.

Disadvantages of Bioreactor Landfills

Bioreactor landfills are a relatively new technology. For the newly developed bioreactor landfills initial monitoring costs are higher to ensure that everything important is discovered and properly controlled. This includes gases, odours and seepage of leachate into the ground surface.

The increased moisture content of bioreactor landfill may reduce the structural stability of the landfill by increasing the pore water pressure within the waste mass.

Since the target of bioreactor landfills is to maintain a high moisture content, gas collection systems can be affected by the increased moisture content of the waste.

Implementation of Bioreactor Landfills

Bioreactor landfills being a novel technology are still in the development phase. Pilot projects for bioreactor landfills are showing promise and more are being experimented with in different parts of the world. Despite the potential benefits of bioreactor landfills there are no standardised and approved designs with guidelines and operational procedures. Following is a list of bioreactor landfill projects which are being used to collect data for forming these needed guidelines and procedures:

United States

- California

  o Yolo County

- Florida

  o Alachua County Southeast Landfill

  o Highlands County

  o New River Regional Landfill, Raiford

  o Polk County Landfill, Winter Haven

- Kentucky

  o Outer Loop Landfill

- Michigan

  o Saint Clair County

- Mississippi

  o Plantation Oaks Bioreactor Demonstration Project, Sibley

- Missouri

  o Columbia

- New Jersey

  o ACUA's Haneman Environmental Park, Egg Harbor Township

- North Carolina

  o Buncombe County Landfill Project

- Virginia

- o   Maplewood Landfill and King George County Landfills
- o   Virginia Landfill Project XL Demonstration Project

Canada

- Sainte-Sophie Bioreactor demonstration Project, Quebec

Australia

- New South Wales

  - o   WoodLawn, Goulburn

- Queensland

  - o   Ti Tree Bioenergy, Ipswich

Landfill Liner

A landfill liner, or composite liner, is intended to be a low a permeable barrier, which is laid down under engineered landfill sites. Until it deteriorates, the liner retards migration of leachate, and its toxic constituents, into underlying aquifers or nearby rivers, causing spoliation of the local water.

A landfill in México showing geomembrane in one of the slopes.

A landfill cell showing a rubberized liner in place on the left.

Modern landfills generally require a layer of compacted clay with a minimum required thickness and a maximum allowable hydraulic conductivity, overlaid by a high-density polyethylene geomembrane.

The United States Environmental Protection Agency has stated that the barriers "will ultimately fail," while the site remains a threat for "thousands of years," suggesting that modern landfill designs delay but do not prevent ground and surface water pollution.

Chipped or waste tires are used to support and insulate the liner.

Mechanical Properties

The primary forms of mechanical degradation associated with geomembranes result from insufficient tensile strength, tear resistance, impact resistance, puncture resistance, and susceptibility to environmental stress cracking (ESC). The ideal method of assessing the amount of liner degradation would be by examining field samples over their service life. Due to the lengths of time required for field sampling tests, various laboratory-accelerated ageing tests have been developed to measure the important mechanical properties.

Tensile Strength

Tensile strength represents the ability for a geomembrane to resist tensile stress. Geomembranes are most commonly tested for tensile strength using one of three methods; the uniaxial tensile test described in ASTM D639-94, the wide-strip tensile test described in ASTM D4885-88, and the multiaxial tension test described in ASTM D5617-94. The difference in these three methods lies in the boundaries imposed into the test specimens. Uniaxial tests do not provide lateral restraint during testing and thus tests the sample under uniaxial stress conditions. During the wide-strip test the sample is restrained laterally while the middle portion is unrestrained. The multiaxial tensile test provides a plane stress boundary condition at the edges of the sample. A typical range of tensile strengths in the machine direction are from 225 to 245 lb/in for 60-mil HDPE to 280 to 325 lb/in for 80-mil HDPE.

Tear Resistance

Tear resistance of a geomembrane becomes important when it is exposed to high winds or handling stress during installation. There are various ASTM methods for measuring tear resistance of geomembranes, with most common reports using ASTM D1004. Typical tear resistances show a value of 40 to 45 lb for 60-mil HDPE and 50 to 60 lb for 80-mil HDPE.

Impact Resistance

Impact resistance provides an assessment of the effects of impacts from falling objects which can either tear or weaken the geomembrane. As with the previous mechanical properties, there are various ASTM methods for assessment. Significantly higher impact resistances are realized when geotextiles are placed above or below the geomembrane. Thicker geomembranes also display higher impact resistances.

Puncture Resistance

Puncture resistance of a geomembrane is important due to the heterogeneous material above and below a typical liner. Rough surfaces, such as stones or other sharp objects, may puncture a membrane if it does not have sufficient puncture resistance. Various methods beyond standard ASTM tests are available; one such method, the critical cone height test, measures the maximum height of a cone on which a compressed geomembrane, which is subjected to increasing pressure, does not fail. HDPE samples typically have a critical cone height of around 1 cm.

Environmental Stress Cracking

Environmental stress cracking is defined as external or internal cracking in plastic induced by applied tensile stress less than its short-term tensile strength. ESC is a fairly common observation in HDPE geomembranes and thus needs to be evaluated carefully. Proper polymeric properties, such as molecular weight, orientation, and distribution, aid in ESC resistance. ASTM D5397 [standard test method for evaluation of stress crack resistance of polyolefin geomembranes using notched constant tensile load (NCTL)] provides the necessary procedure for measuring the ESC resistance of most HDPE geomembranes. The current recommended transition time for an acceptable HDPE geomembrane is around 100 h.

Landfill Mining

Landfill mining and reclamation (LFMR) is a process whereby solid wastes which have previously been landfilled are excavated and processed. The function of landfill mining is to reduce the amount of landfill mass encapsulated within the closed landfill and/or temporarily remove hazardous material to allow protective measures to be taken before the landfill mass is replaced. In the process, mining recovers valuable recyclable materials, a combustible fraction, soil, and landfill space. The aeration of the landfill soil is a secondary benefit regarding the landfill's future use. The combustible fraction is useful for the generation of power. The overall appearance of the landfill mining procedure is a sequence of processing machines laid out in a functional conveyor system. The operating principle is to excavate, sieve and sort the landfill material.

The concept of mining was introduced as early as 1953 at the Hiriya landfill operated by the Dan Region Authority next to the city of Tel Aviv, Israel. Waste contains many resources with high value, the most notable of which are non-ferrous metals such as aluminium cans and scrap metal. The concentration of aluminium in many landfills is higher than the concentration of aluminum in bauxite from which the metal is derived.

Practical Applications

Landfill mining is also possible in countries where land is not available for new landfill sites. In this instance landfill space can be reclaimed by the extraction of biodegradable waste and other substances then refilled with wastes requiring disposal.

Mining construction landfill sites is the simplest form of landfill mining. Construction landfills contain three basic components, wood, scrap metal and gypsum, or drywall, along with a minimal amount of other construction materials. The wood collected can be used as fuel in coal burning power plants and the scrap metal reprocessed.

Mining of municipal landfills is more complicated and has to be based on the expected content of the landfill. Older landfills, in the United States before 1994, were often capped and closed, essentially entombing the waste. This can be beneficial for waste recovery. It can also create a higher risk for toxic waste and leachate exposure as the landfill has not fully processed the stewing wastes. Mining of bioreactor landfills and properly stabilized modern sanitary landfills provides its own benefits. The biodegradable wastes are more easily sieved out, leaving the non-biodegradable materials readily accessible. The quality of these materials for recycling and reprocessing purposes is not as high as initially recycled materials, however materials such as aluminum and steel are usually excluded from this.

Landfill mining is most useful as a method to remediate hazardous landfills. Landfills that were established before landfill liner technology was well established often leak their unprocessed leachate into underlying aquifers. This is both an environmental hazard and also a legal liability. In the US, the Environmental Protection Agency requires closed landfills to be monitored for at least 30 years after waste placement ceases. Mining the landfill simply to lay a safe liner is a last, but sometimes necessary resort.

Methods

The parts of the mining process are the different mining machines. Depending on the complexity of the process more or fewer machines can be used. Machinery is easily transported on trucks from site to site, mounted on trailers.

An excavator or front end loader uncovers the landfilled materials and places them on a moving floor conveyor belt to be taken to the sorting machinery. A trommel is used to separate materials by size. First, a large trommel separates materials like appliances and fabrics. A smaller trommel then allows the biodegraded soil fraction to pass through leaving non-biodegradable, recyclable materials on the screen to be collected.

An electromagnet is used to remove the ferrous material from the waste mass as it passes along the conveyor belt.

A front end loader is used to move sorted materials to trucks for further processing.

Odor control sprayers are wheeled tractors with a cab and movable spray arm mounted on a rotating platform. A large reservoir tank mounted behind the cab holds neutralizing agents, usually in liquid form, to reduce the smell of exposed wastes.

Depending on the level of resource recovery, material can be put through an air classifier which separates light organic material from heavy organic material. The separate streams are then loaded, by front end loaders, onto trucks either for further processing or for sale. Further manual processing can be done on site if processing facilities are too far away to justify the transportation costs.

Incineration

Incineration is a waste treatment process that involves the combustion of organic substances contained in waste materials. Incineration and other high-temperature waste treatment systems are described as "thermal treatment". Incineration of waste materials converts the waste into ash, flue gas, and heat. The ash is mostly formed by the inorganic constituents of the waste, and may take the form of solid lumps or particulates carried by the flue gas. The flue gases must be cleaned of gaseous and particulate pollutants before they are dispersed into the atmosphere. In some cases, the heat generated by incineration can be used to generate electric power.

SYSAV incineration plant in Malmö, Sweden, capable of handling 25 metric tons (28 short tons) per hour of household waste. To the left of the main stack, a new identical oven line is under construction (March 2007).

Incineration with energy recovery is one of several waste-to-energy (WtE) technologies such as gasification, pyrolysis and anaerobic digestion. While incineration and gasification technologies are similar in principle, the energy product from incineration is high-temperature heat whereas combustible gas is often the main energy product from gasification. Incineration and gasification may also be implemented without energy and materials recovery.

In several countries, there are still concerns from experts and local communities about the environmental effect of incinerators.

In some countries, incinerators built just a few decades ago often did not include a materials separation to remove hazardous, bulky or recyclable materials before combustion. These facilities tended to risk the health of the plant workers and the local environment due to inadequate levels of gas cleaning and combustion process control. Most of these facilities did not generate electricity.

Incinerators reduce the solid mass of the original waste by 80–85% and the volume (already compressed somewhat in garbage trucks) by 95–96%, depending on composition and degree of recovery of materials such as metals from the ash for recycling. This means that while incineration does not completely replace landfilling, it significantly reduces the necessary volume for disposal. Garbage trucks often reduce the volume of waste in a built-in compressor before delivery to the incinerator. Alternatively, at landfills, the volume of the uncompressed garbage can be reduced by approximately 70% by using a stationary steel compressor, albeit with a significant energy cost. In many countries, simpler waste compaction is a common practice for compaction at landfills.

Incineration has particularly strong benefits for the treatment of certain waste types in niche areas such as clinical wastes and certain hazardous wastes where pathogens and toxins can be destroyed by high temperatures. Examples include chemical multi-product plants with diverse toxic or very toxic wastewater streams, which cannot be routed to a conventional wastewater treatment plant.

Waste combustion is particularly popular in countries such as Japan where land is a scarce resource. Denmark and Sweden have been leaders in using the energy generated from incineration for more than a century, in localised combined heat and power facilities supporting district heating schemes. In 2005, waste incineration produced 4.8% of the electricity consumption and 13.7% of the total domestic heat consumption in Denmark. A number of other European countries rely heavily on incineration for handling municipal waste, in particular Luxembourg, the Netherlands, Germany, and France.

History

The first UK incinerators for waste disposal were built in Nottingham by Manlove, Alliott & Co. Ltd. in 1874 to a design patented by Albert Fryer. They were originally known as destructors.

The first US incinerator was built in 1885 on Governors Island in New York, NY. The first facility in the Czech Republic was built in 1905 in Brno.

Manlove, Alliott & Co. Ltd. 1894 destructor furnace at Cambridge Museum of Technology

The Encyclopaedia Britannica Eleventh Edition contains a detailed contemporaneous description of the history and design of destructors. For more information, see that entry.

Technology

An incinerator is a furnace for burning waste. Modern incinerators include pollution mitigation equipment such as flue gas cleaning. There are various types of incinerator plant design: moving grate, fixed grate, rotary-kiln, and fluidised bed.

Burn Pile

The burn pile is one of the simplest and earliest forms of waste disposal, essentially consisting of a mound of combustible materials piled on open ground and set on fire.

A typical small burn pile in a garden.

Burn piles can and have spread uncontrolled fires, for example if wind blows burning material off the pile into surrounding combustible grasses or onto buildings. As interior structures of the pile are consumed, the pile can shift and collapse, spreading the burn area. Even in a situation of no wind, small lightweight ignited embers can lift off the pile via convection, and waft through the air into grasses or onto buildings, igniting them.

Burn Barrel

The burn barrel is a somewhat more controlled form of private waste incineration, containing the burning material inside a metal barrel, with a metal grating over the exhaust. The barrel prevents the spread of burning material in windy conditions, and as the combustibles are reduced they can only settle down into the barrel. The exhaust grating helps to prevent the spread of burning embers. Typically steel 55-US-gallon (210 L) drums are used as burn barrels, with air vent holes cut or drilled around the base for air intake. Over time, the very high heat of incineration causes the metal to oxidize and rust, and eventually the barrel itself is consumed by the heat and must be replaced.

Private burning of dry cellulosic/paper products is generally clean-burning, producing no visible smoke, but plastics in household waste can cause private burning to create a public nuisance, generating acrid odors and fumes that make eyes burn and water. Most urban communities ban burn barrels, and certain rural communities may have prohibitions on open burning, especially those home to many residents not familiar with this common rural practice.

As of 2006 in the United States, private rural household or farm waste incineration of small quantities was typically permitted so long as it is not a nuisance to others, does not pose a risk of fire such as in dry conditions, and the fire does not produce dense, noxious smoke. A handful of states, such as New York, Minnesota, and Wisconsin, have laws or regulations either banning or strictly regulating open burning due to health and nuisance effects. People intending to burn waste may be required to contact a state agency in advance to check current fire risk and conditions, and to alert officials of the controlled fire that will occur.

Moving Grate

Control room of a typical moving grate incinerator overseeing two boiler lines

The typical incineration plant for municipal solid waste is a moving grate incinerator. The moving grate enables the movement of waste through the combustion chamber to be optimised to allow a more efficient and complete combustion. A single moving grate boiler can handle up to 35 metric tons (39 short tons) of waste per hour, and can operate 8,000 hours per year with only one scheduled stop for inspection and maintenance of about one month's duration. Moving grate incinerators are sometimes referred to as Municipal Solid Waste Incinerators (MSWIs).

The waste is introduced by a waste crane through the "throat" at one end of the grate, from where it moves down over the descending grate to the ash pit in the other end. Here the ash is removed through a water lock.

Municipal solid waste in the furnace of a moving grate incinerator capable of handling 15 metric tons (17 short tons) of waste per hour. The holes in the grate elements supplying the primary combustion air are visible.

Part of the combustion air (primary combustion air) is supplied through the grate from below. This air flow also has the purpose of cooling the grate itself. Cooling is important for the mechanical strength of the grate, and many moving grates are also water-cooled internally.

Secondary combustion air is supplied into the boiler at high speed through nozzles over the grate. It facilitates complete combustion of the flue gases by introducing turbulence for better mixing and by ensuring a surplus of oxygen. In multiple/stepped hearth incinerators, the secondary combustion air is introduced in a separate chamber downstream the primary combustion chamber.

According to the European Waste Incineration Directive, incineration plants must be designed to ensure that the flue gases reach a temperature of at least 850 °C (1,560 °F) for 2 seconds in order to ensure proper breakdown of toxic organic substances. In order to comply with this at all times, it is required to install backup auxiliary burners (often fueled by oil), which are fired into the boiler in case the heating value of the waste becomes too low to reach this temperature alone.

The flue gases are then cooled in the superheaters, where the heat is transferred to steam, heating the steam to typically 400 °C (752 °F) at a pressure of 40 bars (580 psi) for the electricity generation in the turbine. At this point, the flue gas has a temperature of around 200 °C (392 °F), and is passed to the flue gas cleaning system.

backyard barrels; 26.6% medical waste; 6.3% municipal wastewater treatment sludge; 5.9% municipal waste combustion; 2.9% industrial wood combustion. Thus, the controlled combustion of waste accounted for 41.7% of the total dioxin inventory.

In 1987, before the governmental regulations required the use of emission controls, there was a total of 8,905.1 grams (314.12 oz) Toxic Equivalence (TEQ) of dioxin emissions from US municipal waste combustors. Today, the total emissions from the plants are 83.8 grams (2.96 oz) TEQ annually, a reduction of 99%.

Backyard barrel burning of household and garden wastes, still allowed in some rural areas, generates 580 grams (20 oz) of dioxins annually. Studies conducted by the US-EPA demonstrated that the emissions from just one family using a burn barrel produced more emissions than an incineration plant disposing of 200 metric tons (220 short tons) of waste per day by 1997 and five times that by 2007 due to increased chemicals in household trash and decreased emissions by municipal incinerators using better technology.

However, the same researchers found that their original estimates for the burn barrel were high, and that the incineration plant used for comparison represented a theoretical 'clean' plant rather than any existing facility. Their later studies found that burn barrels produced a median of 24.95 nanograms TEQ per lb garbage burned, so that a family burning 5 lbs of trash per day, or 1825 lbs per year, produces a total of 0.0455 mg TEQ per year, and that the equivalent number of burn barrels for the 83.8 grams (2.96 oz) of the 251 municipal waste combustors inventoried by the EPA in the U.S. in 2000, is 1,841,700, or on average, 7337 family burn barrels per municipal waste incinerator.

Most of the improvement in U.S. dioxin emissions has been for large-scale municipal waste incinerators. As of the year 2000, although small-scale incinerators (those with a daily capacity of less than 250 tons) processed only 9% of the total waste combusted, these produced 83% of the dioxins and furans emitted by municipal waste combustion.

Dioxin Cracking Methods and Limitations

The breakdown of dioxin requires exposure of the molecular ring to a sufficiently high temperature so as to trigger thermal breakdown of the strong molecular bonds holding it together. Small pieces of fly ash may be somewhat thick, and too brief an exposure to high temperature may only degrade dioxin on the surface of the ash. For a large volume air chamber, too brief an exposure may also result in only some of the exhaust gases reaching the full breakdown temperature. For this reason there is also a time element to the temperature exposure to ensure heating completely through the thickness of the fly ash and the volume of waste gases.

There are trade-offs between increasing either the temperature or exposure time. Generally where the molecular breakdown temperature is higher, the exposure time for heating can be shorter, but excessively high temperatures can also cause wear and damage to other parts of the incineration equipment. Likewise the breakdown temperature can be lowered to some degree but then the exhaust gases would require a greater lingering period of perhaps several minutes, which would require large/long treatment chambers that take up a great deal of treatment plant space.

A side effect of breaking the strong molecular bonds of dioxin is the potential for breaking the bonds of nitrogen gas (N_2) and oxygen gas (O_2) in the supply air. As the exhaust flow cools, these highly reactive detached atoms spontaneously reform bonds into reactive oxides such as NO_x in the flue gas, which can result in smog formation and acid rain if they were released directly into the local environment. These reactive oxides must be further neutralized with selective catalytic reduction (SCR) or selective non-catalytic reduction.

Dioxin Cracking in Practice

The temperatures needed to break down dioxin are typically not reached when burning plastics outdoors in a burn barrel or garbage pit, causing high dioxin emissions as mentioned above. While plastic does usually burn in an open-air fire, the dioxins remain after combustion and either float off into the atmosphere, or may remain in the ash where it can be leached down into groundwater when rain falls on the ash pile. Fortunately, dioxin and furan compounds bond very strongly to solid surfaces and are not dissolved by water, so leaching processes are limited to the first few milimeters below the ash pile. The gas-phase dioxins can be substantially destroyed using catalysts, some of which can be present as part of the fabric filter bag structure.

Modern municipal incinerator designs include a high-temperature zone, where the flue gas is sustained at a temperature above 850 °C (1,560 °F) for at least 2 seconds before it is cooled down. They are equipped with auxiliary heaters to ensure this at all times. These are often fueled by oil or natural gas, and are normally only active for a very small fraction of the time. Further, most modern incinerators utilize fabric filters (often with Teflon membranes to enhance collection of sub-micron particles) which can capture dioxins present in or on solid particles.

For very small municipal incinerators, the required temperature for thermal breakdown of dioxin may be reached using a high-temperature electrical heating element, plus a selective catalytic reduction stage.

Although dioxins and furans may be destroyed by combustion, their reformation by a process known as 'de novo synthesis' as the emission gases cool is a probable source of the dioxins measured in emission stack tests from plants that have high combustion temperatures held at long residence times.

CO_2

As for other complete combustion processes, nearly all of the carbon content in the waste is emitted as CO_2 to the atmosphere. MSW contains approximately the same mass fraction of carbon as CO_2 itself (27%), so incineration of 1 ton of MSW produces approximately 1 ton of CO_2.

If the waste was landfilled, 1 ton of MSW would produce approximately 62 cubic metres (2,200 cu ft) methane via the anaerobic decomposition of the biodegradable part of the waste. Since the global warming potential of methane is 34 and the weight of 62 cubic meters of methane at 25 degrees Celsius is 40.7 kg, this is equivalent to 1.38 ton of CO_2, which is more than the 1 ton of CO_2 which would have been produced by incineration. In some countries, large amounts of landfill gas are collected. Still the global warming potential of the landfill gas emitted to atmosphere is significant. In the US it was estimated that the global warming potential of the emitted landfill gas in 1999 was approximately 32% higher than the amount of CO_2 that would have been emitted by incineration. Since this study, the global warming potential estimate for methane has been increased from 21 to 35, which alone would increase this estimate to almost the triple GWP effect compared to incineration of the same waste.

In addition, nearly all biodegradable waste has biological origin. This material has been formed by plants using atmospheric CO_2 typically within the last growing season. If these plants are regrown the CO_2 emitted from their combustion will be taken out from the atmosphere once more.

Such considerations are the main reason why several countries administrate incineration of biodegradable waste as renewable energy. The rest – mainly plastics and other oil and gas derived products – is generally treated as non-renewables.

Different results for the CO_2 footprint of incineration can be reached with different assumptions. Local conditions (such as limited local district heating demand, no fossil fuel generated electricity to replace or high levels of aluminium in the waste stream) can decrease the CO_2 benefits of incineration. The methodology and other assumptions may also influence the results significantly. For example, the methane emissions from landfills occurring at a later date may be neglected or given less weight, or biodegradable waste may not be considered CO_2 neutral. A study by Eunomia Research and Consulting in 2008 on potential waste treatment technologies in London demonstrated that by applying several of these (according to the authors) unusual assumptions the average existing incineration plants performed poorly for CO_2 balance compared to the theoretical potential of other emerging waste treatment technologies.

Other Emissions

Other gaseous emissions in the flue gas from incinerator furnaces include nitrogen oxides, sulfur dioxide, hydrochloric acid, heavy metals, and fine particles. Of the heavy metals, mercury is a major concern due to its toxicity and high volatility, as essentially all mercury in the municipal waste stream may exit in emissions if not removed by emission controls.

The steam content in the flue may produce visible fume from the stack, which can be perceived as a visual pollution. It may be avoided by decreasing the steam content by flue-gas condensation and reheating, or by increasing the flue gas exit temperature well above its dew point. Flue-gas condensation allows the latent heat of vaporization of the water to be recovered, subsequently increasing the thermal efficiency of the plant.

Flue-gas Cleaning

The quantity of pollutants in the flue gas from incineration plants may or may not be reduced by several processes, depending on the plant.

Electrodes inside electrostatic precipitator

Particulate is collected by particle filtration, most often electrostatic precipitators (ESP) and/or baghouse filters. The latter are generally very efficient for collecting fine particles. In an investigation by the Ministry of the Environment of Denmark in 2006, the average particulate emissions per energy content of incinerated waste from 16 Danish incinerators were below 2.02 g/GJ (grams per energy content of the incinerated waste). Detailed measurements of fine particles with sizes below 2.5 micrometres ($PM_{2.5}$) were performed on three of the incinerators: One incinerator equipped with an ESP for particle filtration emitted 5.3 g/GJ fine particles, while two incinerators equipped with baghouse filters emitted 0.002 and 0.013 g/GJ $PM_{2.5}$. For ultra fine particles ($PM_{1.0}$), the numbers were 4.889 g/GJ $PM_{1.0}$ from the ESP plant, while emissions of 0.000 and 0.008 g/GJ $PM_{1.0}$ were measured from the plants equipped with baghouse filters.

Acid gas scrubbers are used to remove hydrochloric acid, nitric acid, hydrofluoric acid, mercury, lead and other heavy metals. The efficiency of removal will depend on the specific equipment, the chemical composition of the waste, the design of the plant, the chemistry of reagents, and the ability of engineers to optimize these conditions, which may conflict for different pollutants. For example, mercury removal by wet scrubbers is considered coincidental and may be less than 50%. Basic scrubbers remove sulfur dioxide, forming gypsum by reaction with lime.

Waste water from scrubbers must subsequently pass through a waste water treatment plant.

Sulfur dioxide may also be removed by dry desulfurisation by injection limestone slurry into the flue gas before the particle filtration.

NO_x is either reduced by catalytic reduction with ammonia in a catalytic converter (selective catalytic reduction, SCR) or by a high-temperature reaction with ammonia in the furnace (selective non-catalytic reduction, SNCR). Urea may be substituted for ammonia as the reducing reagent but must be supplied earlier in the process so that it can hydrolyze into ammonia. Substitution of urea can reduce costs and potential hazards associated with storage of anhydrous ammonia.

Heavy metals are often adsorbed on injected active carbon powder, which is collected by particle filtration.

Solid Outputs

Incineration produces fly ash and bottom ash just as is the case when coal is combusted. The total amount of ash produced by municipal solid waste incineration ranges from 4 to 10% by volume and 15–20% by weight of the original quantity of waste, and the fly ash amounts to about 10–20% of the total ash. The fly ash, by far, constitutes more of a potential health hazard than does the bottom ash because the fly ash often contain high concentrations of heavy metals such as lead, cadmium, copper and zinc as well as small amounts of dioxins and furans. The bottom ash seldom contain significant levels of heavy metals. In testing over the past decade, no ash from an incineration plant in the USA has ever been determined to be a hazardous waste. At present although some historic samples tested by the incinerator operators' group would meet the being eco-toxic criteria at present the EA say "we have agreed" to regard incinerator bottom ash as "non-hazardous" until the testing programme is complete.

Operation of an incinerator aboard an aircraft carrier

Other Pollution Issues

Odor pollution can be a problem with old-style incinerators, but odors and dust are extremely well controlled in newer incineration plants. They receive and store the waste in an enclosed area with a negative pressure with the airflow being routed through the boiler which prevents unpleasant odors from escaping into the atmosphere. However, not all plants are implemented this way, resulting in inconveniences in the locality.

An issue that affects community relationships is the increased road traffic of waste collection vehicles to transport municipal waste to the incinerator. Due to this reason, most incinerators are located in industrial areas. This problem can be avoided to an extent through the transport of waste by rail from transfer stations.

Debate

Use of incinerators for waste management is controversial. The debate over incinerators typically involves business interests (representing both waste generators and incinerator firms), government regulators, environmental activists and local citizens who must weigh the economic appeal of local industrial activity with their concerns over health and environmental risk.

People and organizations professionally involved in this issue include the U.S. Environmental Protection Agency and a great many local and national air quality regulatory agencies worldwide.

Arguments for Incineration

Kehrichtverbrennungsanlage Zürcher Oberland (KEZO) in Hinwil, Switzerland

- The concerns over the health effects of dioxin and furan emissions have been significantly lessened by advances in emission control designs and very stringent new governmental regulations that have resulted in large reductions in the amount of dioxins and furans emissions.

- The U.K. Health Protection Agency concluded in 2009 that "Modern, well managed incinerators make only a small contribution to local concentrations of air pollutants. It is possible that such small additions could have an impact on health but such effects, if they exist, are likely to be very small and not detectable."

- Incineration plants can generate electricity and heat that can substitute power plants powered by other fuels at the regional electric and district heating grid, and steam supply for industrial customers. Incinerators and other waste-to-energy plants generate at least partially biomass-based renewable energy that offsets greenhouse gas pollution from coal-, oil- and gas-fired power plants. The E.U. considers energy generated from biogenic waste (waste with biological origin) by incinerators as non-fossil renewable energy under its emissions caps. These greenhouse gas reductions are in addition to those generated by the avoidance of landfill methane.

- The bottom ash residue remaining after combustion has been shown to be a non-hazardous solid waste that can be safely put into landfills or recycled as construction aggregate. Samples are tested for ecotoxic metals.

- In densely populated areas, finding space for additional landfills is becoming increasingly difficult.

The Maishima waste treatment center in Osaka, designed by Friedensreich Hundertwasser, uses heat for power generation.

Fine particles can be efficiently removed from the flue gases with baghouse filters. Even though approximately 40% of the incinerated waste in Denmark was incinerated at plants with no baghouse filters, estimates based on measurements by the Danish Environmental Research Institute showed that incinerators were only responsible for approximately 0.3% of the total domestic emissions of particulate smaller than 2.5 micrometres ($PM_{2.5}$) to the atmosphere in 2006.

- Incineration of municipal solid waste avoids the release of methane. Every ton of MSW incinerated, prevents about one ton of carbon dioxide equivalents from being released to the atmosphere.

- Most municipalities that operate incineration facilities have higher recycling rates than neighboring cities and counties that do not send their waste to incinerators. This is in part due to enhanced recovery of ceramic materials reused in construction, as well as ferrous and in some cases non-ferrous metals that can be recovered from combustion residue. Metals recovered from ash would typically be difficult or impossible to recycle through conventional means, as the removal of attached combustible material through incineration provides an alternative to labor- or energy-intensive mechanical separation methods.

- Volume of combusted waste is reduced by approximately 90%, increasing the life of landfills. Ash from modern incinerators is vitrified at temperatures of 1,000 °C (1,830 °F) to 1,100 °C (2,010 °F), reducing the leachability and toxicity of residue. As a result, special landfills are generally no longer required for incinerator ash from municipal waste streams, and existing landfills can see their life dramatically increased by combusting waste, reducing the need for municipalities to site and construct new landfills.

Arguments Against Incineration

Decommissioned Kwai Chung Incineration Plant from 1978. It was demolished by February 2009.

- The Scottish Protection Agency's (SEPA) comprehensive health effects research concluded "inconclusively" on health effects in October 2009. The authors stress, that even though no conclusive evidence of non-occupational health effects from incinerators were found in the existing literature, "small but important effects might be virtually impossible to detect". The report highlights epidemiological deficiencies in previous UK health studies and suggests areas for future studies. The U.K. Health Protection Agency produced a lesser summary in September 2009. Many toxicologists criticise and dispute this report as not being comprehensive epidemiologically, thin on peer review and the effects of fine particle effects on health.

- The highly toxic fly ash must be safely disposed of. This usually involves additional waste miles and the need for specialist toxic waste landfill elsewhere. If not done properly, it may cause concerns for local residents.

- The health effects of dioxin and furan emissions from old incinerators; especially during start up and shut down, or where filter bypass is required continue to be a problem.

- Incinerators emit varying levels of heavy metals such as vanadium, manganese, chromium, nickel, arsenic, mercury, lead, and cadmium, which can be toxic at very minute levels.

- Incinerator Bottom Ash (IBA) has elevated levels of heavy metals with ecotoxicity concerns if not reused properly. Some people have the opinion that IBA reuse is still in its infancy and is still not considered to be a mature or desirable product, despite additional engineering treatments. Concerns of IBA use in Foam Concrete have been expressed by the UK Health and Safety Executive in 2010 following several construction and demolition explosions. In its guidance document, IBA is currently banned from use by the UK Highway Authority in concrete work until these incidents have been investigated.

- Alternative technologies are available or in development such as mechanical biological treatment, anaerobic digestion (MBT/AD), autoclaving or mechanical heat treatment (MHT) using steam or plasma arc gasification (PGP), which is incineration using electrically produced extreme high temperatures, or combinations of these treatments.

- Erection of incinerators compete with the development and introduction of other emerging technologies. A UK government WRAP report, August 2008 found that in the UK median incinerator costs per ton were generally higher than those for MBT treatments by £18 per metric ton; and £27 per metric ton for most modern (post 2000) incinerators.

- Building and operating waste processing plants such as incinerators requires long contract periods to recover initial investment costs, causing a long term lock-in. Incinerator lifetimes normally range 25–30 years. This was highlighted by Peter Jones, OBE, the Mayor of London's waste representative in April 2009.

- Incinerators produce fine particles in the furnace. Even with modern particle filtering of the flue gases, a small part of these is emitted to the atmosphere. $PM_{2.5}$ is not separately regulated in the European Waste Incineration Directive, even though they are repeatedly correlated spatially to infant mortality in the UK (M. Ryan's ONS data based maps around the EfW/CHP waste incinerators at Edmonton, Coventry, Chineham, Kirklees and Sheffield). Under WID there is no requirement to monitor stack top or downwind incinerator $PM_{2.5}$ levels. Several European doctors associations (including cross discipline experts such as physicians, environmental chemists and toxicologists) in June 2008 representing over 33,000 doctors wrote a keynote statement directly to the European Parliament citing widespread concerns on incinerator particle emissions and the absence of specific fine and ultrafine particle size monitoring or in depth industry/government epidemiological studies of these minute and invisible incinerator particle size emissions.

- Local communities are often opposed to the idea of locating waste processing plants such as incinerators in their vicinity (the Not in My Back Yard phenomenon). Studies in Andover, Massachusetts correlated 10% property devaluations with close incinerator proximity.

- Prevention, waste minimisation, reuse and recycling of waste should all be preferred to incineration according to the waste hierarchy. Supporters of zero waste consider incinerators and other waste treatment technologies as barriers to recycling and separation beyond particular levels, and that waste resources are sacrificed for energy production.

- A 2008 Eunomia report found that under some circumstances and assumptions, incineration causes less CO_2 reduction than other emerging EfW and CHP technology combinations for treating residual mixed waste. The authors found that CHP incinerator technology without waste recycling ranked 19 out of 24 combinations (where all alternatives to incineration were combined with advanced waste recycling plants); being 228% less efficient than the ranked 1 Advanced MBT maturation technology; or 211% less efficient than plasma gasification/autoclaving combination ranked 2.

- Some incinerators are visually undesirable. In many countries they require a visually intrusive chimney stack.

- If reusable waste fractions are handled in waste processing plants such as incinerators in developing nations, it would cut out viable work for local economies. It is estimated that there are 1 million people making a livelihood off collecting waste.

- The reduced levels of emissions from municipal waste incinerators and waste to energy plants from historical peaks are largely the product of the proficient use of emission control technology. Emission controls add to the initial and operational expenses. It should not be assumed that all new plants will employ the best available control technology if not required by law.

Trends in Incinerator Use

The history of municipal solid waste (MSW) incineration is linked intimately to the history of landfills and other waste treatment technology. The merits of incineration are inevitably judged in relation to the alternatives available. Since the 1970s, recycling and other prevention measures have changed the context for such judgements. Since the 1990s alternative waste treatment technologies have been maturing and becoming viable.

Incineration is a key process in the treatment of hazardous wastes and clinical wastes. It is often imperative that medical waste be subjected to the high temperatures of incineration to destroy pathogens and toxic contamination it contains.

Incineration in North America

The first incinerator in the U.S. was built in 1885 on Governors Island in New York. In 1949, Robert C. Ross founded one of the first hazardous waste management companies in the U.S. He began Robert Ross Industrial Disposal because he saw an opportunity to meet the hazardous waste management needs of companies in northern Ohio. In 1958, the company built one of the first hazardous waste incinerators in the U.S.

The first full-scale, municipally operated incineration facility in the U.S. was the Arnold O. Chantland Resource Recovery Plant, built in 1975 and located in Ames, Iowa. This plant is still in operation and produces refuse-derived fuel that is sent to local power plants for fuel. The first commercially successful incineration plant in the U.S. was built in Saugus, Massachusetts in October 1975 by Wheelabrator Technologies, and is still in operation today.

There are several environmental or waste management corporations that transport ultimately to an incinerator or cement kiln treatment center. Currently (2009), there are three main businesses that incinerate waste: Clean Harbours, WTI-Heritage, and Ross Incineration Services. Clean Harbours has acquired many of the smaller, independently run facilities, accumulating 5–7 incinerators in the process across the U.S. WTI-Heritage has one incinerator, located in the southeastern corner of Ohio across the Ohio River from West Virginia.

Several old generation incinerators have been closed; of the 186 MSW incinerators in 1990, only 89 remained by 2007, and of the 6200 medical waste incinerators in 1988, only 115 remained in 2003. No new incinerators were built between 1996 and 2007. The main reasons for lack of activity have been:

- Economics. With the increase in the number of large inexpensive regional landfills and, up until recently, the relatively low price of electricity, incinerators were not able to compete for the 'fuel', i.e., waste in the U.S.

- Tax policies. Tax credits for plants producing electricity from waste were rescinded in the U.S. between 1990 and 2004.

There has been renewed interest in incineration and other waste-to-energy technologies in the U.S. and Canada. In the U.S., incineration was granted qualification for renewable energy production tax credits in 2004. Projects to add capacity to existing plants are underway, and municipalities are once again evaluating the option of building incineration plants rather than continue landfilling municipal wastes. However, many of these projects have faced continued political opposition in spite of renewed arguments for the greenhouse gas benefits of incineration and improved air pollution control and ash recycling.

Incineration in Europe

In Europe, with the ban on landfilling untreated waste, scores of incinerators have been built in the last decade, with more under construction. Recently, a number of municipal governments have begun the process of contracting for the construction and operation of incinerators. In Europe, some of the electricity generated from waste is deemed to be from a 'Renewable Energy Source (RES) and is thus eligible for tax credits if privately operated. Also, some incinerators in Europe are equipped with waste recovery, allowing the reuse of ferrous and non-ferrous materials found in landfills. A prominent example is the AEB Waste Fired Power Plant.

In Sweden, about 50% of the generated waste is burned in waste-to-energy facilities, producing electricity and supplying local cities' district heating systems. The importance of waste in Sweden's electricity generation scheme is reflected on their 700.000 tons of waste imported per year to supply waste-to-energy facilities.

Incineration in the United Kingdom

The technology employed in the UK waste management industry has been greatly lagging behind that of Europe due to the wide availability of landfills. The Landfill Directive set down by the European Union led to the Government of the United Kingdom imposing waste legislation including the landfill tax and Landfill Allowance Trading Scheme. This legislation is designed to reduce the release of greenhouse gases produced by landfills through the use of alternative methods of waste treatment. It is the UK Government's position that incineration will play an increasingly large role in the treatment of municipal waste and supply of energy in the UK.

In 2008, plans for potential incinerator locations exists for approximately 100 sites. These have been interactively mapped by UK NGO's.

Under a new plan in June 2012, a DEFRA-backed grant scheme (The Farming and Forestry Improvement Scheme) was set up to encourage the use of low-capacity incinerators on agricultural sites to improve their bio security.

Incineration Units for Emergency Use

Emergency incineration systems exist for the urgent and biosecure disposal of animals and their by-products following a mass mortality or disease outbreak. An increase in regulation and enforcement from governments and institutions worldwide has been forced through public pressure and significant economic exposure.

Contagious animal disease has cost governments and industry $200 billion over 20 years to 2012 and is responsible for over 65% of infectious disease outbreaks worldwide in the past sixty years. One-third of global meat exports (approx 6 million tonnes) is affected by trade restrictions at any time and as such the focus of Governments, public bodies and commercial operators is on cleaner, safer and more robust methods of animal carcass disposal to contain and control disease.

Mobile incineration unit for emergency use

Large-scale incineration systems are available from niche suppliers and are often bought by governments as a safety net in case of contagious outbreak. Many are mobile and can be quickly deployed to locations requiring biosecure disposal.

Small Incinerator Units

Small-scale incinerators exist for special purposes. For example, the small-scale incinerators are aimed for hygienically safe destruction of medical waste in developing countries. Small incinerators can be quickly deployed to remote areas where an outbreak has occurred to dispose of infected animals quickly and without the risk of cross contamination.

An example of a low capacity, mobile incinerator

In Popular Media

- In Cube Zero, fictional so-called "flash Incinerators" exist, which essentially vaporize anything organic.

- Incinerators make an appearance in SimCity 3000 in two varieties: a large, traditional combustion device that spews out a significant amount of air pollution, and a more modern device that converts the waste into energy to power the city with a bigger capacity to load the garbage, though still producing a lot of pollution.

- They also make an appearance in SimCity 4, but without the non-energy-from-waste variant.

- In the climax of Portal (video game), the main protagonist, Chell, while on a conveyor belt, escapes an incinerator, after the game's main antagonist, GLaDOS, forced her into it.

- The climax of Toy Story 3 features an infamous scene, where the working of a moving-grate incinerator (and of a garbage shredder) was shown dramatically from the inside, as the toys face destruction.

Compost

Compost is organic matter that has been decomposed and recycled as a fertilizer and soil amendment. Compost is a key ingredient in organic farming.

A community-level composting plant in a rural area in Germany

At the simplest level, the process of composting simply requires making a heap of wetted organic matter known as green waste (leaves, food waste) and waiting for the materials to break down into humus after a period of weeks or months. Modern, methodical composting is a multi-step, closely monitored process with measured inputs of water, air, and carbon- and nitrogen-rich materials. The decomposition process is aided by shredding the plant matter, adding water and ensuring proper aeration by regularly turning the mixture. Worms and fungi further break up the material. Bacteria requiring oxygen to function (aerobic bacteria) and fungi manage the chemical process by converting the inputs into heat, carbon dioxide and ammonium. The ammonium (NH_4) is the form of nitrogen used by plants. When available ammonium is not used by plants it is further converted by bacteria into nitrates (NO_3) through the process of nitrification.

Compost is rich in nutrients. It is used in gardens, landscaping, horticulture, and agriculture. The compost itself is beneficial for the land in many ways, including as a soil conditioner, a fertilizer, addition of vital humus or humic acids, and as a natural pesticide for soil. In ecosystems, compost is useful for erosion control, land and stream reclamation, wetland construction, and as landfill cover. Organic ingredients intended for composting can alternatively be used to generate biogas through anaerobic digestion.

Terminology

Composting of waste is an aerobic (in the presence of air) method of decomposing solid wastes. The process involves decomposition of organic waste into humus known as compost which is a good fertiliser for plants. However, the term "composting" is used worldwide with differing meanings. Some composting textbooks narrowly define composting as being an aerobic form of decomposition, primarily by microbes. An alternative term to composting is "aerobic digestion", which in turn is also referred to as "wet composting".

For many people, composting is used to refer to several different types of biological process. In North America, "anaerobic composting" is still a common term for what much of the rest of the world and in technical publications people call "anaerobic digestion". The microbes used and the processes involved are quite different between composting and anaerobic digestion.

Ingredients

Home compost barrel in the Escuela Barreales, Santa Cruz, Chile

Carbon, Nitrogen, Oxygen, Water

Materials in a compost pile

Composting organisms require four equally important ingredients to work effectively:

- Carbon — for energy; the microbial oxidation of carbon produces the heat, if included at suggested levels.

  o High carbon materials tend to be brown and dry.

- Nitrogen — to grow and reproduce more organisms to oxidize the carbon.

  o High nitrogen materials tend to be green (or colorful, such as fruits and vegetables) and wet.

- Oxygen — for oxidizing the carbon, the decomposition process.

- Water — in the right amounts to maintain activity without causing anaerobic conditions.

Certain ratios of these materials will provide beneficial bacteria with the nutrients to work at a rate that will heat up the pile. In that process much water will be released as vapor ("steam"), and the oxygen will be quickly depleted, explaining the need to actively manage the pile. The hotter the pile gets, the more often added air and water is necessary; the air/water balance is critical to maintaining high temperatures (135°-160° Fahrenheit / 50° - 70° Celsius) until the materials are broken down. At the same time, too much air or water also slows the process, as does too much carbon (or too little nitrogen). Hot container composting focuses on retaining the heat to increase decomposition rate and produce compost quicker.

The most efficient composting occurs with an optimal carbon:nitrogen ratio of about 10:1 to 20:1. Rapid composting is favored by having a C/N ratio of ~30 or less. Theoretical analysis is confirmed by field tests that above 30 the substrate is nitrogen starved, below 15 it is likely to outgas a portion of nitrogen as ammonia. If nitrogen needs to be increased, it has been suggested to add 0.15 pounds of *actual* nitrogen per three bushels (3.75 cubic feet) of lower nitrogen material. [For those not familiar with these types of units: 0.64g/L or 640 grams of actual nitrogen per cubic meter.] Two to 3 pounds of organic nitrogen supplement (blood meal, manure, bone meal, alfalfa meal) per 100 pounds of low nitrogen materials (for example, straw or sawdust), supplies generally ample nitrogen and trace minerals in high carbon mixes.

Food scraps compost heap

Nearly all plant and animal materials have both carbon and nitrogen, but amounts vary widely, with characteristics noted above (dry/wet, brown/green). Fresh grass clippings have an average ratio of about 15:1 and dry autumn leaves about 50:1 depending on species. Mixing equal parts by volume approximates the ideal C:N range. Few individual situations will provide the ideal mix of materials at any point. Observation of amounts, and consideration of different materials as a pile is built over time, can quickly achieve a workable technique for the individual situation.

Animal Manure and Bedding

On many farms, the basic composting ingredients are animal manure generated on the farm and bedding. Straw and sawdust are common bedding materials. Non-traditional bedding materials are also used, including newspaper and chopped cardboard. The amount of manure composted on a livestock farm is often determined by cleaning schedules, land availability, and weather conditions. Each type of manure has its own physical, chemical, and biological characteristics. Cattle and horse manures, when mixed with bedding, possess good qualities for composting. Swine manure, which is very wet and usually not mixed with bedding material, must be mixed with straw or similar raw materials. Poultry manure also must be blended with carbonaceous materials - those low in nitrogen preferred, such as sawdust or straw.

Microorganisms

With the proper mixture of water, oxygen, carbon, and nitrogen, micro-organisms are allowed to break down organic matter to produce compost. The composting process is dependent on micro-organisms to break down organic matter into compost. There are many types of microorganisms found in active compost of which the most common are:

- Bacteria- The most numerous of all the microorganisms found in compost. Depending on the phase of composting, mesophilic or thermophilic bacteria may predominate.

- Actinobacteria- Necessary for breaking down paper products such as newspaper, bark, etc.

- Fungi- Molds and yeast help break down materials that bacteria cannot, especially lignin in woody material.

- Protozoa- Help consume bacteria, fungi and micro organic particulates.

- Rotifers- Rotifers help control populations of bacteria and small protozoans.

In addition, earthworms not only ingest partly composted material, but also continually re-create aeration and drainage tunnels as they move through the compost.

A lack of a healthy micro-organism community is the main reason why composting processes are slow in landfills with environmental factors such as lack of oxygen, nutrients or water being the cause of the depleted biological community.

Phases of Composting

Under ideal conditions, composting proceeds through three major phases:

- An initial, mesophilic phase, in which the decomposition is carried out under moderate temperatures by mesophilic microorganisms.

- As the temperature rises, a second, thermophilic phase starts, in which the decomposition is carried out by various thermophilic bacteria under high temperatures.

- As the supply of high-energy compounds dwindles, the temperature starts to decrease, and the mesophiles once again predominate in the maturation phase.

Human Waste

Human waste (excreta) can also be added as an input to the composting process, like it is done in composting toilets, as human waste is a nitrogen-rich organic material.

People excrete far more water-soluble plant nutrients (nitrogen, phosphorus, potassium) in urine than in feces. Human urine can be used directly as fertilizer or it can be put onto compost. Adding a healthy person's urine to compost usually will increase temperatures and therefore increase its ability to destroy pathogens and unwanted seeds. Urine from a person with no obvious symptoms of infection is much more sanitary than fresh feces. Unlike feces, urine does not attract disease-spreading flies (such as house flies or blow flies), and it does not contain the most hardy of pathogens, such as parasitic worm eggs. Urine usually does not stink for long, particularly when it is fresh, diluted, or put on sorbents.

Urine is primarily composed of water and urea. Although metabolites of urea are nitrogen fertilizers, it is easy to over-fertilize with urine, or to utilize urine containing pharmaceutical (or other) content, creating too much ammonia for plants to absorb, acidic conditions, or other phytotoxicity.

Humanure

"Humanure" is a portmanteau of *human* and *manure*, designating human excrement (feces and urine) that is recycled via composting for agricultural or other purposes. The term was first used in a 1994 book by Joseph Jenkins that advocates the use of this organic soil amendment. The term humanure is used by compost enthusiasts in the US but not generally elsewhere. Because the term "humanure" has no authoritative definition it is subject to various uses; news reporters occasionally fail to correctly distinguish between humanure and sewage sludge or "biosolids".

Uses

Compost is generally recommended as an additive to soil, or other matrices such as coir and peat, as a tilth improver, supplying humus and nutrients. It provides a rich *growing medium*, or a porous, absorbent material that holds moisture and soluble minerals, providing the support and nutrients in which plants can flourish, although it is rarely used alone, being primarily mixed with soil, sand, grit, bark chips, vermiculite, perlite, or clay granules to produce loam. Compost can be tilled directly into the soil or growing medium to boost the level of organic matter and the overall fertility of the soil. Compost that is ready to be used as an additive is dark brown or even black with an earthy smell.

Generally, direct seeding into a compost is not recommended due to the speed with which it may dry and the possible presence of phytotoxins that may inhibit germination, and the possible tie up of nitrogen by incompletely decomposed lignin. It is very common to see blends of 20–30% compost used for transplanting seedlings at cotyledon stage or later.

Composting can destroy pathogens or unwanted seeds. Unwanted living plants (or weeds) can be discouraged by covering with mulch/compost. The "microbial pesticides" in compost may include thermophiles and mesophiles, however certain composting detritivores such as black soldier fly larvae and redworms, also reduce many pathogens. Thermophilic (high-temperature) composting is well known to destroy many seeds and nearly all types of pathogens (exceptions may include prions). The sanitizing qualities of (thermophilic) composting are desirable where there is a high likelihood of pathogens, such as with manure.

Composting Technologies

A homemade compost tumbler

A modern compost bin constructed from plastics

Overview

In addition to the traditional compost pile, various approaches have been developed to handle different composting processes, ingredients, locations, and applications for the composted product.

There is a large number of different composting systems on the market, for example:

- At the household level: Composting toilet, container composting, vermicomposting

- At the industrial composting (large scale): Aerated Static Pile Composting, vermicomposting, windrow composting etc.

Examples

Vermicomposting

Vermicompost is the product or process of composting through the utilization of various species of worms, usually red wigglers, white worms, and earthworms, to create a heterogeneous mixture of decomposing vegetable or food waste (excluding meat, dairy, fats, or oils), bedding materials, and vermicast. Vermicast, also known as worm castings, worm humus or worm manure, is the end-product of the breakdown of organic matter by species of earthworm. Vermicomposting is widely used in North America for on-site institutional processing of food waste, such as in hospitals and shopping malls. This type of composting is sometimes suggested as a feasible indoor home composting method. Vermicomposting has gained popularity in both these industrial and domestic settings because, as compared with conventional composting, it provides a way to compost organic materials more quickly (as defined by a higher rate of carbon-to-nitrogen ratio increase) and to attain products that have lower salinity levels that are therefore more beneficial to plant mediums.

Rotary screen harvested worm castings

The earthworm species (or composting worms) most often used are red wigglers (*Eisenia fetida* or *Eisenia andrei*), though European nightcrawlers (*Eisenia hortensis* or *Dendrobaena veneta*) could also be used. Red wigglers are recommended by most vermiculture experts, as they have some of the best appetites and breed very quickly. Users refer to European nightcrawlers by a variety of other names, including *dendrobaenas*, *dendras*, Dutch Nightcrawlers, and Belgian nightcrawlers.

Food waste - after three years

Containing water-soluble nutrients, vermicompost is a nutrient-rich organic fertilizer and soil conditioner in a form that is relatively easy for plants to absorb. Worm castings are sometimes used as an organic fertilizer. Because the earthworms grind and uniformly mix minerals in simple forms, plants need only minimal effort to obtain them.

The worms' digestive systems also add beneficial microbes to help create a "living" soil environment for plants.

Vermicompost tea in conjunction with 10% castings has been shown to cause up to a 1.7 times growth in plant mass over plants grown without.

Researchers from the Pondicherry University discovered that worm composts can also be used to clean up heavy metals. The researchers found substantial reductions in heavy metals when the worms were released into the garbage and they are effective at removing lead, zinc, cadmium, copper and manganese.

Hügelkultur (Raised Garden Beds or Mounds)

The practice of making raised garden beds or mounds filled with rotting wood is also called "Hügelkultur" in German. It is in effect creating a Nurse log that is covered with dirt.

An almost completed Hügelkultur bed; the bed does not have dirt on it yet.

Benefits of hügelkultur garden beds include water retention and warming of soil. Buried wood becomes like a sponge as it decomposes, able to capture water and store it for later use by crops planted on top of the hügelkultur bed.

The buried decomposing wood will also give off heat, as all compost does, for several years. These effects have been used by Sepp Holzer to enable fruit trees to survive at otherwise inhospitable temperatures and altitudes.

Black Soldier Fly Larvae Composting

Black Soldier Fly (*Hermetia illucens*) larvae have been shown to be able to rapidly consume large amounts of organic waste when kept at 31.8 °C, the optimum temperature for reproduction. Enthusiasts have experimented with a large number of different waste products and some even sell starter kits to the public.

Cockroach Composting

Cockroach composting is another insect-mediated composting method. In this case the adults of any number of cockroach species (such as the Turkestan cockroach or *Blaptica dubia*) are used to quickly convert manure or kitchen waste to nutrient dense compost. Depending on species used and environmental conditions, excess composting insects can be used as an excellent animal feed for farm animals and pets.

Bokashi

Bokashi is a method that uses a mix of microorganisms to cover food waste or wilted plants to decrease smell. Bokashi (ぼかし) is Japanese for "shading off" or "gradation." It derives from the practice of Japanese farmers centuries ago of covering food waste with rich, local soil that contained the microorganisms that would ferment the waste. After a few weeks, they would bury the waste.

Inside a recently started bokashi bin. The aerated base is just visible through the food scraps and bokashi bran.

Most practitioners obtain the microorganisms from the product Effective Microorganisms (EM1), first sold in the 1980s. EM1 is mixed with a carbon base (e.g. sawdust or bran) that it sticks to and a sugar for food (e.g. molasses). The mixture is layered with waste in a sealed container and after a few weeks, removed and buried.

Newspaper fermented in a lactobacillus culture can be substituted for bokashi bran for a successful bokashi bucket.

Compost Tea

Compost teas are defined as water extracts brewed from composted materials and can be derived from aerobic or anaerobic processes. Compost teas are generally produced from adding one volume of compost to 4-10 volumes of water, but there has also been debate about the benefits of aerating the mixture. Field studies have shown the benefits of adding compost teas to crops due to the adding of organic matter, increased nutrient availability and increased microbial activity. They have also been shown to have an effect on plant pathogens.

Composting Toilets

A composting toilet does not require water or electricity, and when properly managed does not smell. A composting toilet collects human excreta which is then added to a compost heap together with sawdust and straw or other carbon rich materials, where pathogens are destroyed to some extent. The amount of pathogen destruction depends on the temperature (mesophilic or thermophilic conditions) and composting time. A composting toilet tries to process the excreta in situ although this is often coupled with a secondary external composting step. The resulting compost product has been given various names, such as humanure and EcoHumus.

A composting toilet can aid in the conservation of fresh water by avoiding the usage of potable water required by the typical flush toilet. It further prevents the pollution of ground water by controlling the fecal matter decomposition before entering the system. When properly managed, there should be no ground contamination from leachate.

Compost and Land-filling

As concern about landfill space increases, worldwide interest in recycling by means of composting is growing, since composting is a process for converting decomposable organic materials into useful stable products. Composting is one of the only ways to revitalize soil vitality due to phosphorus depletion in soil. Industrial scale composting in the form of in-vessel composting, aerated static pile composting, and anaerobic digestion takes place in most Western countries now, and in many areas is mandated by law. There are process and product guidelines in Europe that date to the early 1980s (Germany, the Netherlands, Switzerland) and only more recently in the UK and the US. In both these countries, private trade associations within the industry have established loose standards, some say as a stop-gap measure to discourage independent government agencies from establishing tougher consumer-friendly standards. The USA is the only Western country that does not distinguish sludge-source compost from green-composts, and by default in the USA 50% of states expect composts to comply in some manner with the federal EPA 503 rule promulgated in 1984 for sludge products. Compost is regulated in Canada and Australia as well.

Industrial Systems

Industrial composting systems are increasingly being installed as a waste management alternative to landfills, along with other advanced waste processing systems. Mechanical sorting of mixed waste streams combined with anaerobic digestion or in-vessel composting is called mechanical biological treatment, and is increasingly being used in developed countries due to regulations controlling the amount of organic matter allowed in landfills. Treating biodegradable waste before it enters a landfill reduces global warming from fugitive methane; untreated waste breaks down anaerobically in a landfill, producing landfill gas that contains methane, a potent greenhouse gas.

Vermicomposting, also known as vermiculture, is used for medium-scale on-site institutional composting, such as for food waste from universities and shopping malls. It is selected either as a more environmentally friendly choice than conventional methods of disposal, or to reduce the cost of commercial waste removal.

A large compost pile that is steaming with the heat generated by thermophilic microorganisms.

Large-scale composting systems are used by many urban areas around the world. Co-composting is a technique that combines solid waste with de-watered biosolids, although difficulties controlling inert and plastics contamination from municipal solid waste makes this approach less attractive. The world's largest MSW co-composter is the Edmonton Composting Facility in Edmonton, Alberta, Canada, which turns 220,000 tonnes of residential solid waste and 22,500 dry tonnes of biosolids per year into 80,000 tonnes of compost. The facility is 38,690 m² (416,500 sq.ft.) in area, equivalent to 4½ Canadian football fields, and the operating structure is the largest stainless steel building in North America, the size of 14 NHL rinks. In 2006, Qatar awarded Keppel Seghers Singapore, a subsidiary of Keppel Corporation, a contract to begin construction on a 275,000 tonne/year anaerobic digestion and composting plant licensed by Kompogas (de) Switzerland. This plant, with 15 independent anaerobic digesters, will be the world's largest composting facility once fully operational in early 2011 and forms part of Qatar's Domestic Solid Waste Management Centre, the largest integrated waste management complex in the Middle East.

Another large MSW composter is the Lahore Composting Facility in Lahore, Pakistan, which has a capacity to convert 1,000 tonnes of municipal solid waste per day into compost. It also has a capacity to convert substantial portion of the intake into refuse-derived fuel (RDF) materials for further combustion use in several energy consuming industries across Pakistan, for example in cement manufacturing companies where it is used to heat cement kilns. This project has also been approved by the Executive Board of the United Nations Framework Convention on Climate Change for reducing methane emissions, and has been registered with a capacity of reducing 108,686 tonnes carbon dioxide equivalent per annum.

Related Technologies

Anaerobic digestion is process for converting organic waste into (biogas). The residual material, sometimes in combination with sewage sludge can be followed by an aerobic composting process before selling or giving away the compost.

History

Composting as a recognized practice dates to at least the early Roman Empire since Pliny the Elder (AD 23-79). Traditionally, composting involved piling organic materials until the next planting season, at which time the materials would have decayed enough to be ready for use in the soil. The advantage of this method is that little working time or effort is required from the composter and it fits in naturally with agricultural practices in temperate climates. Disadvantages (from the modern perspective) are that space is used for a whole year, some nutrients might be leached due to exposure to rainfall, and disease-producing organisms and insects may not be adequately controlled.

Compost Basket

Composting was somewhat modernized beginning in the 1920s in Europe as a tool for organic farming. The first industrial station for the transformation of urban organic materials into compost was set up in Wels, Austria in the year 1921. Early frequent citations for propounding composting within farming are for the German-speaking world Rudolf Steiner, founder of a farming method called biodynamics, and Annie Francé-Harrar, who was appointed on behalf of the government in Mexico and supported the country 1950–1958 to set up a large humus organization in the fight against erosion and soil degradation.

In the English-speaking world it was Sir Albert Howard who worked extensively in India on sustainable practices and Lady Eve Balfour who was a huge proponent of composting. Composting was imported to America by various followers of these early European movements by the likes of J.I. Rodale (founder of Rodale Organic Gardening), E.E. Pfeiffer (who developed scientific practices in biodynamic farming), Paul Keene (founder of Walnut Acres in Pennsylvania), and Scott and Helen Nearing (who inspired the back-to-the-land movement of the 1960s). Coincidentally, some of the above met briefly in India - all were quite influential in the U.S. from the 1960s into the 1980s.

There are many modern proponents of rapid composting that attempt to correct some of the perceived problems associated with traditional, slow composting. Many advocate that compost can be made in 2 to 3 weeks. Many such short processes involve a few changes to traditional methods, including smaller, more homogenized pieces in the compost, controlling carbon-to-nitrogen ratio (C:N) at 30 to 1 or less, and monitoring the moisture level more carefully. However, none of these parameters differ significantly from the early writings of Howard and Balfour, suggesting that in fact modern composting has not made significant advances over the traditional methods that take a few months to work. For this reason and others, many modern scientists who deal with carbon transformations are sceptical that there is a "super-charged" way to get nature to make compost rapidly.

In fact, both sides are right to some extent. The bacterial activity in rapid high heat methods breaks down the material to the extent that pathogens and seeds are destroyed, and the original feedstock is unrecognizable. At this stage, the compost can be used to prepare fields or other planting areas. However, most professionals recommend that the compost be given time to cure before using in a nursery for starting seeds or growing young plants. The curing time allows fungi to continue the decomposition process and eliminating phytotoxic substances.

Many countries such as Wales and some individual cities such as Seattle and San Francisco require food and yard waste to be sorted for composting.

Kew Gardens in London has one of the biggest non-commercial compost heaps in Europe.

Related Lists

- List of composting systems

- List of environment topics

- List of sustainable agriculture topics

- List of organic gardening and farming topics

References

- Hickmann, H. Lanier, Jr. (2003). American alchemy: the history of solid waste management in the United States. ForesterPress. ISBN 978-0-9707687-2-8.

- Cleveland, Cutler J.; Morris, Christopher G. (November 15, 2013). Handbook of Energy: Chronologies, Top Ten Lists, and Word Clouds. Elsevier. p. 461. ISBN 978-0-12-417019-3.

- Carl A. Zimring (2005). Cash for Your Trash: Scrap Recycling in America. New Brunswick, NJ: Rutgers University Press. ISBN 0-8135-4694-X.

- Lynn R. Kahle; Eda Gurel-Atay, eds. (2014). Communicating Sustainability for the Green Economy. New York: M.E. Sharpe. ISBN 978-0-7656-3680-5.

- Huesemann, M.; Huesemann, J. (2011). Techno-fix: Why Technology Won't Save Us or the Environment. New Society Publishers. p. 464. ISBN 978-0-86571-704-6. Retrieved 2016-07-07.

- Foster, J. B.; Clark, B. (2011). The Ecological Rift: Capitalisms War on the Earth. Monthly Review Press. p. 544. ISBN 1-58367-218-4.

- Steven E. Landsburg. "Why I Am Not An Environmentalist: The Science of Economics Versus the Religion of Ecology Excerpt from The Armchair Economist: Economics & Everyday Life" (PDF) (PDF). Retrieved July 6, 2016.

- "Report: "On the Making of Silk Purses from Sows' Ears," 1921: Exhibits: Institute Archives & Special Collections: MIT". mit.edu. Retrieved July 7, 2016.

- "Cost Effective Waste to Energy Technologies – Updated Article With Extra Information". bionomicfuel.com. Retrieved 28 February 2015.

- "Autonomie énergétique pour un refuge de montagne : panneaux solaires". Connaissance des Énergies. 5 July 2012. Retrieved 28 February 2015.

- "Bulgaria opens largest WEEE recycling factory in Eastern Europe". www.ask-eu.com. WtERT Germany GmbH. July 12, 2010. Retrieved July 29, 2015.

- "EnvironCom opens largest WEEE recycling facility / waste & recycling news". www.greenwise-business.co.uk. The Sixty Mile Publishing Company. March 4, 2010. Retrieved July 29, 2015.

- Goodman, Peter S. (January 11, 2012). "Where Gadgets Go To Die: E-Waste Recycler Opens New Plant In Las Vegas". The Huffington Post. Retrieved July 29, 2015.

- Moses, Asher (November 19, 2008). "New plant tackles our electronic leftovers – BizTech – Technology – smh.com.au". www.smh.com.au. Retrieved July 29, 2015.

- "A Beverage Container Deposit Law for Hawaii". www.opala.org. City & County of Honolulu, Department of Environmental Services. Oct 2002. Retrieved July 31, 2015.

Alternative Strategies of Waste Management

A number of strategies are used to manage waste. Some of the strategies are co-processing, curb mining and land farming. Waste that can be used as raw material or as a source of energy is referred to as co-processing. These strategies teach the recycling and usage of waste in whichever manner possible, to save resources and energy.

Co-processing

Co-processing is the use of waste as raw material, or as a source of energy, or both to replace natural mineral resources (material recycling) and fossil fuels such as coal, petroleum and gas (energy recovery) in industrial processes, mainly in energy intensive industries (EII) such as cement, lime, steel, glass, and power generation. Waste materials used for Co-processing are referred to as alternative fuels and raw materials (AFR).

Concept of Co-processing

Co-processing is a proven sustainable development concept that reduces demands on natural resources, reduces pollution and landfill space, thus contributing to reducing the environmental footprint. Co-processing is also based on the principles of industrial ecology, which considers the best features of the flow of information, materials, and energy of biological ecosystems, with the aim of improving the exchange of these essential resources in the industrial world.

Waste		Substitution	Examples
Energy content (carbon; hydrogen)	Energy recovery	Substitution of fossil energy	Solvents Waste oil Waste plastics
Material content (CaO, Fe₂O₃, Al₂O₃, etc.)	Material recovery	Substitution of raw material	Used tires Used paints Industrial sludge
Energy content (carbon; hydrogen)	Energy recovery	Substitution of fossil energy	
Material content (CaO, Fe₂O₃, Al₂O₃, etc.)	Material recovery	Substitution of raw material	Molding sand Blast furnace slag Fly ash & bottom ash By-product gypsum

Figure 1: Types of Co-processing

In summary, the benefits of Co-processing are:

- to conserve natural (non-renewable) resources of energy and materials,

- to reduce emissions of greenhouse gases in order to slow global warming and demonstrate a positive impact on integrated environmental indicators, such as the ecological footprint,

- to reduce the environmental impacts of the extraction (mining or quarrying), transporting, and processing of raw materials,

- to reduce dependence on primary resource markets,

- to save landfill space and reduce the pollution caused by the disposal of waste, and

- to destroy waste completely eliminating potential future liabilities.

Co-processing contributes to the industrial competitiveness, is a complementary technology to concepts such as cleaner production or recycling and should be considered as a treatment alternative within an integrated waste management concept. Some EII offer co-processing as a sustainable waste management service. It is usually more cost effective to adapt existing facilities of EII than building new waste treatment capacities thereby reducing waste management cost to society.

The waste management hierarchy shows that Co-processing is a recovery activity which should be considered after waste prevention and recycling; Co-processing ranks higher in this hierarchy in comparison to disposal activities such as landfilling or incineration.

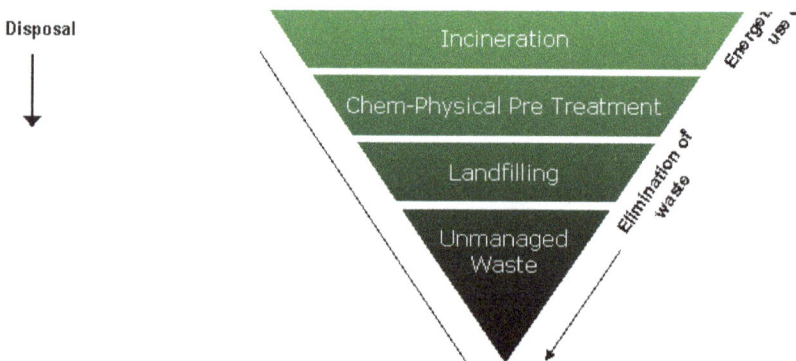

Figure 2: Waste Management Hierarchy

Potential of Co-processing

The global industrial demand for energy is roughly 45% of the total demand and the requirements of the energy intensive industries (EII) are more than half of the total industrial demand, at 27%.

Worldwide, wastes suitable for Co-processing have an energy potential equivalent to nearly 20% of the fossil fuel energy used by the EII and coal-fired power plants. By 2030, the thermal substitution rate of waste could rise to nearly 30%. In the EU-25 countries of Europe, the available energy potential in waste currently represents nearly 40% of this demand, and this is expected to rise to almost 50% by 2030.

However, in the year 2004 in EU-25, less than 10% of the energy content of the waste that was not being reused or recycled was utilized by the EII and power plants. This figure indicates to which extend the high potential of waste as alternative fuel and a source of materials is being neglected.

Roughly 60% of the waste that could be used for Co-processing is biomass and therefore carbon neutral. In this way Co-processing offers a significant potential for the reduction of greenhouse gas emissions from fossil fuels. Furthermore, diverting industrial waste streams from landfills and incinerators without energy recovery contributes to reducing overall CO_2 emissions when used to substitute fossil fuels through Co-processing (as illustrated in the figure below).

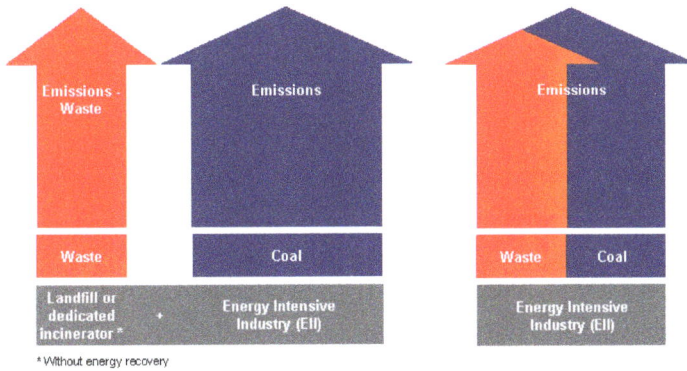

Figure 3: Reduction of Emissions through Co-processing

Other factors that must be considered when Co-processing waste include product quality standards, permitting aspects, and transparent communication in order to gain public acceptance.

Examples of Waste that Can be Co-processed

Used tires

Various solid waste in plastic barrels

Shredded solid waste

Curb Mining

Curb mining is the act of finding furniture and art discarded on the street ("curbside"). In cities around the world, people often dispose of furniture and other unwanted items by leaving them on the sidewalk for others to take.

Terms similar to curb mining include "dumpster diving" and "freeganism". In June 2007, The New York Times wrote:

"Freegans" are scavengers of the developed world, living off consumer waste in an effort to minimize their support of corporations and their impact on the planet, and to distance themselves from what they see as out-of-control consumerism. They forage through supermarket trash and eat the slightly-bruised produce or just-expired canned goods that are routinely thrown out, and negotiate gifts of surplus food from sympathetic stores and restaurants. They dress in castoff clothes and furnish their homes with items found on the street.

—Steven Kurutz,

Re-use and Recycling

In many jurisdictions, ownership of domestic waste changes once it's placed into a container for collection. It's thus illegal (although rarely enforced) to skip dive. Curb mining gets round this because the items offered are not yet placed (in a legal sense) into the 'waste' stream, thus their ownership hasn't yet been transferred. It's often legal to curb mine, but illegal to skip dive.

Some countries, notably Germany, Japan and much of Western mainland Europe, have a long tradition that items placed outside are intended specifically for re-use by others. There may be a designated day of the week or month, distinct from normal refuse collections, to encourage this.

Marketing

The urban phenomenon of curb mining has been used by various companies for experimental marketing. The strategy is to create awareness of a product by handing it out for free.

In 2006, Tom Dixon, a London designer, handed out 500 of his polystyrene chairs to a crowd in Trafalgar Square. The next year, in the same location, he gave away 1,000 energy-efficient light bulbs that he had designed.

In 2009, advertising agency Mono and modern furniture designer Blu Dot created an experiment to see what would happen if they left 25 Blu Dot chairs on the street for "curb miners" to find. They attached GPS devices to the chairs, which were activated once the chairs were picked up and taken. The chairs were then tracked back to the new owners' homes where a handful of them were interviewed for a documentary.

Landfarming

Landfarming is an ex-situ waste treatment process that is performed in the upper soil zone or in biotreatment cells. Contaminated soils, sediments, or sludges are transported to the landfarming site, incorporated into the soil surface and periodically turned over (tilled) to aerate the mixture. Landfarming commonly uses a clay or composite liner to intercept leaching contaminants and prevent groundwater pollution, however, a liner is not a universal requirement.

Applicability

This technique has been used for years in the management and disposal of drill cuttings, oily sludge and other petroleum refinery wastes. The equipment employed in land farming is typical of that used in agricultural operations. These land farming activities cultivate and enhance microbial degradation of hazardous compounds. As a rule of thumb, the higher the molecular weight (i.e., the more rings within a polycyclic aromatic hydrocarbon), the slower the degradation rate. Also, the more chlorinated or nitrated the compound, the more difficult it is to degrade.

Limitations

Factors that may limit the applicability and effectiveness of the process include:

1. large space requirements

2. the conditions advantageous for biological degradation of contaminants are largely uncontrolled, which increases the required length of time until complete degradation, particularly for recalcitrant compounds

3. inorganic contaminants are not biodegraded

4. the potential of large amounts of particulate matter released by operations

5. the presence of metal ions may be toxic to microbes and may leach from the contaminated soil into the ground.

Hydrocarbon compounds that have been identified as being not readily degraded by land farming include creosote, pentachlorophenol (PCP), and bunker C oil.

References

- Best Practice Note: Landfarming (PDF), Sydney NSW: Environment Protection Authority, April 2014, ISBN 978-1-74359-607-4

- Dennis R. Heldman, ed. (2003), Encyclopedia of Agricultural, Food, and Biological Engineering, CRC Press, p. 114, ISBN 978-0824709389.

Wastewater Treatment Technologies

Water that is gravely affected by human activities is known as wastewater. A common example of wastewater is sewage. A number of techniques have been used to treat sewage. Some of these are aerobic treatment system, trickling filter, sewage treatment and industrial wastewater treatment. This chapter is an overview of the subject matter incorporating all the major aspects of waste management.

Aerobic Treatment System

An aerobic treatment system or ATS, often called (incorrectly) an aerobic septic system, is a small scale sewage treatment system similar to a septic tank system, but which uses an aerobic process for digestion rather than just the anaerobic process used in septic systems. These systems are commonly found in rural areas where public sewers are not available, and may be used for a single residence or for a small group of homes.

Unlike the traditional septic system, the aerobic treatment system produces a high quality secondary effluent, which can be sterilized and used for surface irrigation. This allows much greater flexibility in the placement of the leach field, as well as cutting the required size of the leach field by as much as half.

Process

The ATS process generally consists of the following phases:

- Pre-treatment stage to remove large solids and other undesirable substances.
- Aeration stage, where aerobic bacteria digest biological wastes.
- Settling stage allows undigested solids to settle. This forms a sludge that must be periodically removed from the system.
- Disinfecting stage, where chlorine or similar disinfectant is mixed with the water, to produce an antiseptic output.

The disinfecting stage is optional, and is used where a sterile effluent is required, such as cases where the effluent is distributed above ground. The disinfectant typically used is tablets of calcium hypochlorite, which are specially made for waste treatment systems. The tablets are intended to break down quickly in sunlight. Stabilized forms of chlorine persist after the effluent is dispersed, and can kill plants in the leach field.

Since the ATS contains a living ecosystem of microbes to digest the waste products in the water, excessive amounts of items such as bleach or antibiotics can damage the ATS environment and reduce treatment effectiveness. Non-digestible items should also be avoided, as they will build up in the system and require more frequent sludge removal.

Types of Aerobic Treatment Systems

Small scale aerobic systems generally use one of two designs, fixed-film systems, or continuous flow, suspended growth aerobic systems (CFSGAS). The pre-treatment and effluent handling are similar for both types of systems, and the difference lies in the aeration stage.

Fixed Film Systems

Fixed film systems use a porous medium which provides a bed to support the biomass film that digests the waste material in the wastewater. Designs for fixed film systems vary widely, but fall into two basic categories (though some systems may combine both methods). The first is a system where the media is moved relative to the wastewater, alternately immersing the film and exposing it to air, while the second uses a stationary media, and varies the wastewater flow so the film is alternately submerged and exposed to air. In both cases, the biomass must be exposed to both wastewater and air for the aerobic digestion to occur. The film itself may be made of any suitable porous material, such as formed plastic or peat moss. Simple systems use stationary media, and rely on intermittent, gravity driven wastewater flow to provide periodic exposure to air and wastewater. A common moving media system is the rotating biological contactor (RBC), which uses disks rotating slowly on a horizontal shaft. Approximately 40 percent of the disks are submerged at any given time, and the shaft rotates at a rate of one or two revolutions per minute.

Continuous Flow, Suspended Growth Aerobic Systems

CFSGAS systems, as the name implies, are designed to handle continuous flow, and do not provide a bed for a bacterial film, relying rather on bacteria suspended in the wastewater. The suspension and aeration are typically provided by an air pump, which pumps air through the aeration chamber, providing a constant stirring of the wastewater in addition to the oxygenation. A medium to promote fixed film bacterial growth may be added to some systems designed to handle higher than normal levels of biomass in the wastewater.

Retrofit or Portable Aerobic Systems

Another increasingly common use of aerobic treatment is for the remediation of failing or failed anaerobic septic systems, by retrofitting an existing system with an aerobic feature. This class of product, known as aerobic remediation, is designed to remediate biologically failed and failing anaerobic distribution systems by significantly reducing the biochemical oxygen demand (BOD5) and total suspended solids (TSS) of the effluent. The reduction of the BOD5 and TSS reverses the developed bio-mat. Further, effluent with high dissolved oxygen and aerobic bacteria flow to the distribution component and digest the bio-mat.

Composting Toilets

Composting toilets are designed to treat only toilet waste, rather than general residential waste water, and are typically used with water-free toilets rather than the flush toilets associated with the above types of aerobic treatment systems. These systems treat the waste as a moist solid, rather than in liquid suspension, and therefore separate urine from feces during treatment to maintain the correct moisture content in the system. An example of a composting toilet is the clivus multrum (Latin for 'inclined chamber'), which consists of an inclined chamber that separates urine and feces and a fan to provide positive ventilation and prevent odors from escaping through the toilet. Within the chamber, the urine and feces are independently broken down not only by aerobic bacteria, but also by fungi, arthropods, and earthworms. Treatment times are very long, with a minimum time between removals of solid waste of a year; during treatment the volume of the solid waste is decreased by 90 percent, with most being converted into water vapor and carbon dioxide. Pathogens are eliminated from the waste by the long durations in inhospitable conditions in the treatment chamber.

Comparison to Traditional Septic Systems

The aeration stage and the disinfecting stage are the primary differences from a traditional septic system; in fact, an aerobic treatment system can be used as a secondary treatment for septic tank effluent. These stages increase the initial cost of the aerobic system, and also the maintenance requirements over the passive septic system. Unlike many other biofilters, aerobic treatment systems require a constant supply of electricity to drive the air pump increasing overall system costs. The disinfectant tablets must be periodically replaced, as well as the electrical components (air compressor) and mechanical components (air diffusers). On the positive side, an aerobic system produces a higher quality effluent than a septic tank, and thus the leach field can be smaller than that of a conventional septic system, and the output can be discharged in areas too environmentally sensitive for septic system output. Some aerobic systems recycle the effluent through a sprinkler system, using it to water the lawn where regulations approve.

Effluent Quality

Since the effluent from an ATS is often discharged onto the surface of the leach field, the quality is very important. A typical ATS will, when operating correctly, produce an effluent with less than 30 mg/liter BOD5, 25 mg/L TSS, and 10,000 cfu/mL fecal coliform bacteria. This is clean enough that it cannot support a biomat or "slime" layer like a septic tank.

ATS effluent is relatively odorless; a properly operating system will produce effluent that smells musty, but not like sewage. Aerobic treatment is so effective at reducing odors, that it is the preferred method for reducing odor from manure produced by farms.

Trickling Filter

A trickling filter is a type of wastewater treatment system first used by Dibden and Clowes It consists of a fixed bed of rocks, lava, coke, gravel, slag, polyurethane foam, sphagnum peat moss, ceramic, or plastic media over which sewage or other wastewater flows downward and causes a layer of microbial slime (biofilm) to grow, covering the bed of media. Aerobic conditions are maintained by splashing, diffusion, and either by forced-air flowing through the bed or natural convection of air if the filter medium is porous.

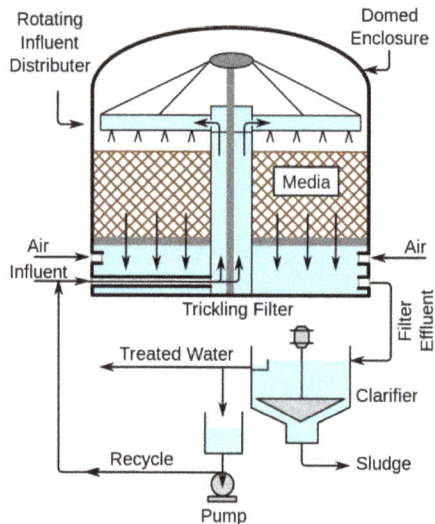

A typical complete trickling filter system

The terms trickle filter, trickling biofilter, biofilter, biological filter and biological trickling filter are often used to refer to a trickling filter. These systems have also been described as roughing filters, intermittent filters, packed media bed filters, alternative septic systems, percolating filters, attached growth processes, and fixed film processes.

Construction

A typical trickling filter is circular and between 10 metres and 20 metres across and between 2 metres to 3 metres deep. A circular wall, often of brick, contains a bed of filter media which in turn rests on a base of under-drains. These under-drains function both to remove liquid passing through the filter media but also to allow the free passage of air up through the filter media. Mounted in the center over the top of the filter media is a spindle supporting two or more horizontal perforated pipes which extend to the edge of the media. The perforations on the pipes are designed to allow an even flow of liquid over the whole area of the media and are also angled so that when liquid flows from the pipes the whole assembly rotates around the central spindle. Settled sewage is delivered to a reservoir at the centre of the spindle via some form of dosing mechanism, often a tipping bucket device on small filters.

Larger filters may be rectangular and the distribution arms may be driven by hydraulic or electrical systems.

Trickling may have a variety of types of filter media used to support the bioi-film. Types of media most commonly used include coke, pumice, plastic matrix material, open-cell polyurethane foam, clinker, gravel, sand and geotextiles. Ideal filter medium optimizes surface area for microbial attachment, wastewater retention time, allows air flow, resists plugging is mechanically robust in all weathers allowing walking access across the filter and does not degrade. Some residential systems require forced aeration units which will increase maintenance and operational costs.

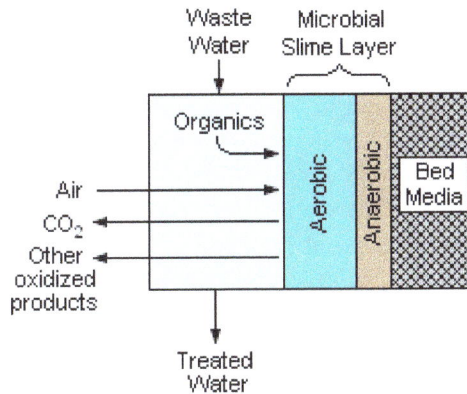

Image 1. A schematic cross-section of the contact face of the bed of media in a trickling filter

Operation

Typically, sewage flow enters at a high level and flows through the primary settlement tank. The supernatant from the tank flows into a dosing device, often a tipping bucket which delivers flow to the arms of the filter. The flush of water flows through the arms and exits through a series of holes pointing at an angle downwards. This propels the arms around distributing the liquid evenly over the surface of the filter media. Most are uncovered (unlike the accompanying diagram) and are freely ventilated to the atmosphere.

Systems can be configured for single-pass use where the treated water is applied to the trickling filter once before being disposed of, or for multi-pass use where a portion of the treated water is cycled back and re-treated via a closed loop. Multi-pass systems result in higher treatment quality and assist in removing Total Nitrogen (TN) levels by promoting nitrification in the aerobic media bed and denitrification in the anaerobic septic tank. Some systems use the filters in two banks operated in series so that the wastewater has two passes through a filter with a sedimentation stage between the two passes. Every few days the filters are switched round to balance the load. This method of treatment can improve nitrification and de-nitrification since much of the carbonaceous oxidative material is removed on the first pass through the filters.

The removal of pollutants from the waste water stream involves both absorption and adsorption of organic compounds and some inorganic species such as nitrite and nitrate ions by the layer of microbial bio film. The filter media is typically chosen to provide a very high surface area to volume. Typical materials are often porous and have considerable internal surface area in addition to the external surface of the medium. Passage of the waste water over the media provides dissolved oxygen which the bio-film layer requires for the biochemical oxidation of the organic compounds and releases carbon dioxide gas, water and other oxidized end products. As the bio film layer thickens, it eventually sloughs off into the liquid flow and subsequently forms part of the secondary sludge. Typically, a trickling filter is followed by a clarifier or sedimentation tank for the separation and removal of the sloughed film. Other filters utilizing higher-density media such as sand, foam and peat moss do not produce a sludge that must be removed, but require forced air blowers and backwashing or an enclosed anaerobic environment.

Biological Processes

The bio-film that develops in a trickling filter may become several millimetres thick and is typically a gelatinous matrix that contains many species of bacteria, cilliates and amoeboid protozoa, annelids, round worms and insect larvae and many other micro fauna. This is very different from many other bio-films which may be less than 1 mm thick. Within the thickness of the biofilm both aerobic and anaerobic zones can exist supporting both oxidative and reductive biological processes. At certain times of year, especially in the spring, rapid growth of organisms in the film may cause the film to be too thick and it may slough off in patches leading to the "spring slough".

Types

Single trickling filters may be used for the treatment of small residential septic tank discharges and very small rural sewage treatment systems. Larger centralized sewage treatment plants typically use many trickling filters in parallel. The treatment of industrial wastewater may involve specialised tricking filters which use plastic media and high flow rates.

Industrial Wastewater Treatment

Wastewaters from a variety of industrial processes have been treated in trickling filters. Such industrial wastewater trickling filters consist of two types:

- Large tanks or concrete enclosures filled with plastic packing or other media.

- Vertical towers filled with plastic packing or other media.

The availability of inexpensive plastic tower packings has led to their use as trickling filter beds in tall towers, some as high as 20 meters. As early as the 1960s, such towers were in use at: the Great Northern Oil's Pine Bend Refinery in Minnesota; the Cities Service Oil Company Trafalgar Refinery in Oakville, Ontario and at a kraft paper mill.

The treated water effluent from industrial wastewater trickling filters is typically processed in a clarifier to remove the sludge that sloughs off the microbial slime layer attached to the trickling filter media as for other trickling filter applications.

Some of the latest trickle filter technology involves aerated biofilters of plastic media in vessels using blowers to inject air at the bottom of the vessels, with either downflow or upflow of the wastewater.

Sewage Treatment

Sewage treatment is the process of removing contaminants from wastewater, primarily from household sewage. It includes physical, chemical, and biological processes to remove these contaminants and produce environmentally safe treated wastewater (or treated effluent). A by-product of sewage treatment is usually a semi-solid waste or slurry, called sewage sludge, that has to undergo further treatment before being suitable for disposal or land application.

Wastewater treatment plant in Massachusetts, United States

Sewage treatment may also be referred to as wastewater treatment, although the latter is a broader term which can also be applied to purely industrial wastewater. For most cities, the sewer system will also carry a proportion of industrial effluent to the sewage treatment plant which has usually received pretreatment at the factories themselves to reduce the pollutant load. If the sewer system is a combined sewer then it will also carry urban runoff (stormwater) to the sewage treatment plant.

Terminology

The term "sewage treatment plant" (or "sewage treatment works" in some countries) is nowadays often replaced with the term "wastewater treatment plant".

Sewage can be treated close to where the sewage is created, which may be called a "decentralized" system or even an "on-site" system (in septic tanks, biofilters or aerobic treatment systems). Alternatively, sewage can be collected and transported by a network of pipes and pump stations to a municipal treatment plant. This is called a "centralized" system, although the borders between decentralized and centralized can be variable. For this reason, the terms "semi-decentralized" and "semi-centralized" are also being used.

Origins of Sewage

Sewage is generated by residential, institutional, commercial and industrial establishments. It includes household waste liquid from toilets, baths, showers, kitchens, and sinks draining into sewers. In many areas, sewage also includes liquid waste from industry and commerce. The separation and draining of household waste into greywater and blackwater is becoming more common in the developed world, with treated greywater being permitted to be used for watering plants or recycled for flushing toilets.

Sewage Mixing With Rainwater

Sewage may include stormwater runoff or urban runoff. Sewerage systems capable of handling storm water are known as combined sewer systems. This design was common when urban sewerage systems were first developed, in the late 19th and early 20th centuries. Combined sewers require much larger and more expensive treatment facilities than sanitary sewers. Heavy volumes of storm runoff may overwhelm the sewage treatment system, causing a spill or overflow. Sanitary sewers are typically much smaller than combined sewers, and they are not designed to transport stormwater. Backups of raw sewage can occur if excessive infiltration/inflow (dilution by stormwater and/or groundwater) is allowed into a sanitary sewer system. Communities that have urbanized in the mid-20th century or later generally have built separate systems for sewage (sanitary sewers) and stormwater, because precipitation causes widely varying flows, reducing sewage treatment plant efficiency.

As rainfall travels over roofs and the ground, it may pick up various contaminants including soil particles and other sediment, heavy metals, organic compounds, animal waste, and oil and grease. Some jurisdictions require stormwater to receive some level of treatment before being discharged directly into waterways. Examples of treatment processes used for stormwater include retention basins, wetlands, buried vaults with various kinds of media filters, and vortex separators (to remove coarse solids).

Industrial Effluent

In highly regulated developed countries, industrial effluent usually receives at least pretreatment if not full treatment at the factories themselves to reduce the pollutant load, before discharge to the sewer. This process is called industrial wastewater treatment. The same does not apply to many developing countries where industrial effluent is more likely to enter the sewer if it exists, or even the receiving water body, without pretreatment.

Industrial wastewater may contain pollutants which cannot be removed by conventional sewage treatment. Also, variable flow of industrial waste associated with production cycles may upset the population dynamics of biological treatment units, such as the activated sludge process.

Process Steps

Overview

Sewage collection and treatment is typically subject to local, state and federal regulations and standards.

Treating wastewater has the aim to produce an effluent that will do as little harm as possible when discharged to the surrounding environment, thereby preventing pollution compared to releasing untreated wastewater into the environment.

Sewage treatment generally involves three stages, called primary, secondary and tertiary treatment.

- *Primary treatment* consists of temporarily holding the sewage in a quiescent basin where heavy solids can settle to the bottom while oil, grease and lighter solids float to the surface. The settled and floating materials are removed and the remaining liquid may be discharged or subjected to secondary treatment. Some sewage treatment plants that are connected to a combined sewer system have a bypass arrangement after the primary treatment unit. This means that during very heavy rainfall events, the secondary and tertiary treatment systems can be bypassed to protect them from hydraulic overloading, and the mixture of sewage and stormwater only receives primary treatment.

- *Secondary treatment* removes dissolved and suspended biological matter. Secondary treatment is typically performed by indigenous, water-borne micro-organisms in a managed habitat. Secondary treatment may require a separation process to remove the micro-organisms from the treated water prior to discharge or tertiary treatment.

- *Tertiary treatment* is sometimes defined as anything more than primary and secondary treatment in order to allow rejection into a highly sensitive or fragile ecosystem (estuaries, low-flow rivers, coral reefs,...). Treated water is sometimes disinfected chemically or physically (for example, by lagoons and microfiltration) prior to discharge into a stream, river, bay, lagoon or wetland, or it can be used for the irrigation of a golf course, green way or park. If it is sufficiently clean, it can also be used for groundwater recharge or agricultural purposes.

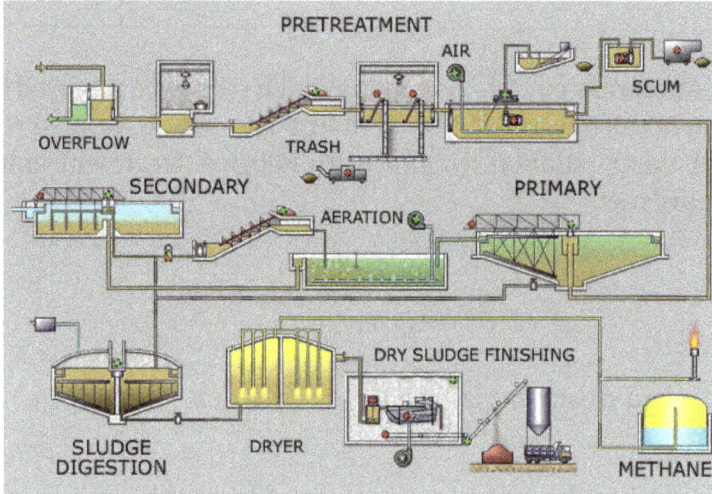

Simplified process flow diagram for a typical large-scale treatment plant

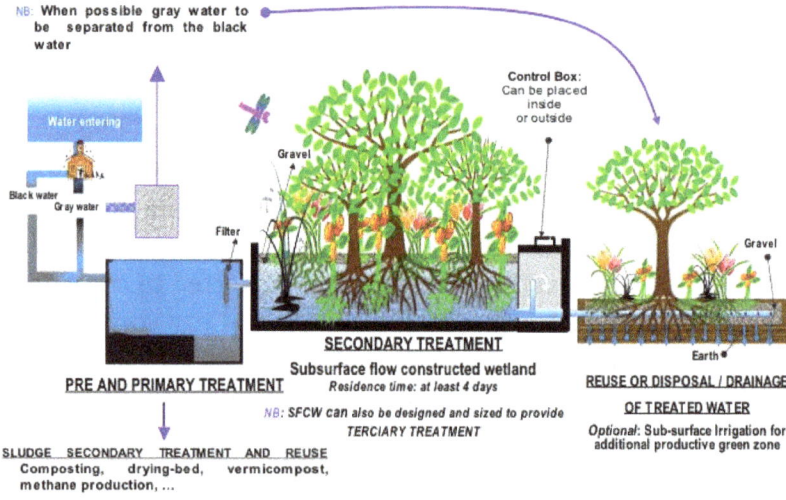

Process flow diagram for a typical treatment plant via subsurface flow constructed wetlands (SFCW)

Pretreatment

Pretreatment removes all materials that can be easily collected from the raw sewage before they damage or clog the pumps and sewage lines of primary treatment clarifiers. Objects commonly removed during pretreatment include trash, tree limbs, leaves, branches, and other large objects.

The influent in sewage water passes through a bar screen to remove all large objects like cans, rags, sticks, plastic packets etc. carried in the sewage stream. This is most commonly done with an automated mechanically raked bar screen in modern plants serving large populations, while in smaller or less modern plants, a manually cleaned screen may be used. The raking action of a mechanical bar screen is typically paced according to the accumulation on the bar screens and/or flow rate. The solids are collected and later disposed in a landfill, or incinerated. Bar screens or mesh screens of varying sizes may be used to optimize solids removal. If gross solids are not removed, they become entrained in pipes and moving parts of the treatment plant, and can cause substantial damage and inefficiency in the process.

Grit Removal

Pretreatment may include a sand or grit channel or chamber, where the velocity of the incoming sewage is adjusted to allow the settlement of sand, grit, stones, and broken glass. These particles are removed because they may damage pumps and other equipment. For small sanitary sewer systems, the grit chambers may not be necessary, but grit removal is desirable at larger plants. Grit chambers come in 3 types: horizontal grit chambers, aerated grit chambers and vortex grit chambers. The process is called sedimentation.

Flow Equalization

Clarifiers and mechanized secondary treatment are more efficient under uniform flow conditions. Equalization basins may be used for temporary storage of diurnal or wet-weather flow peaks. Basins provide a place to temporarily hold incoming sewage during plant maintenance and a means of diluting and distributing batch discharges of toxic or high-strength waste which might otherwise inhibit biological secondary treatment (including portable toilet waste, vehicle holding tanks, and septic tank pumpers). Flow equalization basins require variable discharge control, typically include provisions for bypass and cleaning, and may also include aerators. Cleaning may be easier if the basin is downstream of screening and grit removal.

Fat and Grease Removal

In some larger plants, fat and grease are removed by passing the sewage through a small tank where skimmers collect the fat floating on the surface. Air blowers in the base of the tank may also be used to help recover the fat as a froth. Many plants, however, use primary clarifiers with mechanical surface skimmers for fat and grease removal.

Primary Treatment

In the primary sedimentation stage, sewage flows through large tanks, commonly called "pre-settling basins", "primary sedimentation tanks" or "primary clarifiers". The tanks are used to settle sludge while grease and oils rise to the surface and are skimmed off. Primary settling tanks are usually equipped with mechanically driven scrapers that continually drive the collected sludge towards a hopper in the base of the tank where it is pumped to sludge treatment facilities. Grease and oil from the floating material can sometimes be recovered for saponification (soap making).

Primary treatment tanks in Oregon, USA.

Secondary Treatment

Secondary treatment is designed to substantially degrade the biological content of the sewage which are derived from human waste, food waste, soaps and detergent. The majority of municipal plants treat the settled sewage liquor using aerobic biological processes. To be effective, the biota require both oxygen and food to live. The bacteria and protozoa consume biodegradable soluble organic contaminants (e.g. sugars, fats, organic short-chain carbon molecules, etc.) and bind much of the less soluble fractions into floc. Secondary treatment systems are classified as *fixed-film* or *suspended-growth* systems.

- Fixed-film or attached growth systems include trickling filters, bio-towers, and rotating biological contactors, where the biomass grows on media and the sewage passes over its surface. The fixed-film principle has further developed into Moving Bed Biofilm Reactors (MBBR) and Integrated Fixed-Film Activated Sludge (IFAS) processes. An MBBR system typically requires a smaller footprint than suspended-growth systems.

- Suspended-growth systems include activated sludge, where the biomass is mixed with the sewage and can be operated in a smaller space than trickling filters that treat the same amount of water. However, fixed-film systems are more able to cope with drastic changes in the amount of biological material and can provide higher removal rates for organic material and suspended solids than suspended growth systems.

Secondary Sedimentation

Some secondary treatment methods include a secondary clarifier to settle out and separate biological floc or filter material grown in the secondary treatment bioreactor.

Secondary clarifier at a rural treatment plant.

List of Process Types

- Activated sludge

- Aerated lagoon

- Aerobic granulation

- Constructed wetland

- Membrane bioreactor

- Rotating biological contactor

- Sequencing batch reactor

- Trickling filter

To use less space, treat difficult waste, and intermittent flows, a number of designs of hybrid treatment plants have been produced. Such plants often combine at least two stages of the three main treatment stages into one combined stage. In the UK, where a large number of wastewater treatment plants serve small populations, package plants are a viable alternative to building a large structure for each process stage. In the US, package plants are typically used in rural areas, highway rest stops and trailer parks.

Tertiary Treatment

The purpose of tertiary treatment is to provide a final treatment stage to further improve the effluent quality before it is discharged to the receiving environment (sea, river, lake, wet lands, ground, etc.). More than one tertiary treatment process may be used at any treatment plant. If disinfection is practised, it is always the final process. It is also called "effluent polishing."

Filtration

Sand filtration removes much of the residual suspended matter. Filtration over activated carbon, also called *carbon adsorption,* removes residual toxins.

Lagoons or Ponds

Lagoons or ponds provide settlement and further biological improvement through storage in large man-made ponds or lagoons. These lagoons are highly aerobic and colonization by native macrophytes, especially reeds, is often encouraged. Small filter feeding invertebrates such as *Daphnia* and species of *Rotifera* greatly assist in treatment by removing fine particulates.

A sewage treatment plant and lagoon in Everett, Washington, United States.

Biological Nutrient Removal

Biological nutrient removal (BNR) is regarded by some as a type of secondary treatment process, and by others as a tertiary (or "advanced") treatment process.

Wastewater may contain high levels of the nutrients nitrogen and phosphorus. Excessive release to the environment can lead to a buildup of nutrients, called eutrophication, which can in turn encourage the overgrowth of weeds, algae, and cyanobacteria (blue-green algae). This may cause an algal bloom, a rapid growth in the population of algae. The algae numbers are unsustainable and eventually most of them die. The decomposition of the algae by bacteria uses up so much of the oxygen in the water that most or all of the animals die, which creates more organic matter for the bacteria to decompose. In addition to causing deoxygenation, some algal species produce toxins that contaminate drinking water supplies. Different treatment processes are required to remove nitrogen and phosphorus.

Nitrogen Removal

Nitrogen is removed through the biological oxidation of nitrogen from ammonia to nitrate (nitrification), followed by denitrification, the reduction of nitrate to nitrogen gas. Nitrogen gas is released to the atmosphere and thus removed from the water.

Nitrification itself is a two-step aerobic process, each step facilitated by a different type of bacteria. The oxidation of ammonia (NH_3) to nitrite (NO_2^-) is most often facilitated by *Nitrosomonas* spp. ("nitroso" referring to the formation of a nitroso functional group). Nitrite oxidation to nitrate (NO_3^-), though traditionally believed to be facilitated by *Nitrobacter* spp. (nitro referring the formation of a nitro functional group), is now known to be facilitated in the environment almost exclusively by *Nitrospira* spp.

Denitrification requires anoxic conditions to encourage the appropriate biological communities to form. It is facilitated by a wide diversity of bacteria. Sand filters, lagooning and reed beds can all be used to reduce nitrogen, but the activated sludge process (if designed well) can do the job the most easily. Since denitrification is the reduction of nitrate to dinitrogen (molecular nitrogen) gas, an electron donor is needed. This can be, depending on the waste water, organic matter (from feces), sulfide, or an added donor like methanol. The sludge in the anoxic tanks (denitrification tanks) must be mixed well (mixture of recirculated mixed liquor, return activated sludge [RAS], and raw influent) e.g. by using submersible mixers in order to achieve the desired denitrification.

Sometimes the conversion of toxic ammonia to nitrate alone is referred to as tertiary treatment.

Over time, different treatment configurations have evolved as denitrification has become more sophisticated. An initial scheme, the Ludzack-Ettinger Process, placed an anoxic treatment zone before the aeration tank and clarifier, using the return activated sludge (RAS) from the clarifier as a nitrate source. Influent wastewater (either raw or as effluent from primary clarification) serves as the electron source for the facultative bacteria to metabolize carbon, using the inorganic nitrate as a source of oxygen instead of dissolved molecular oxygen. This denitrification scheme was naturally limited to the amount of soluble nitrate present in the RAS. Nitrate reduction was limited because RAS rate is limited by the performance of the clarifier.

The "Modified Ludzak-Ettinger Process" (MLE) is an improvement on the original concept, for it recycles mixed liquor from the discharge end of the aeration tank to the head of the anoxic tank to provide a consistent source of soluble nitrate for the facultative bacteria. In this instance, raw wastewater continues to provide the electron source, and sub-surface mixing maintains the bacteria in contact with both electron source and soluble nitrate in the absence of dissolved oxygen.

Many sewage treatment plants use centrifugal pumps to transfer the nitrified mixed liquor from the aeration zone to the anoxic zone for denitrification. These pumps are often referred to as *Internal Mixed Liquor Recycle* (IMLR) pumps. IMLR may be 200% to 400% the flow rate of influent wastewater (Q.) This is in addition to Return Activated Sludge (RAS) from secondary clarifiers, which may be 100% of Q. (Therefore, the hydraulic capacity of the tanks in such a system should handle at least 400% of annual average design flow (AADF.) At times, the raw or primary effluent wastewater must be carbon-supplemented by the addition of methanol, acetate, or simple food waste (molasses, whey, plant starch) to improve the treatment efficiency. These carbon additions should be accounted for in the design of a treatment facility's organic loading.

Further modifications to the MLE were to come: Bardenpho and Biodenipho processes include additional anoxic and oxidative processes to further polish the conversion of nitrate ion to molecular nitrogen gas. Use of an anaerobic tank following the initial anoxic process allows for luxury uptake of phosphorus by bacteria, thereby biologically reducing orthophosphate ion in the treated wastewater. Even newer improvements, such as Anammox Process, interrupt the formation of nitrate at the nitrite stage of nitrification, shunting nitrite-rich mixed liquor activated sludge to treatment where nitrite is then converted to molecular nitrogen gas, saving energy, alkalinity, and secondary carbon sourcing. Anammox™ (ANaerobic AMMonia OXidation) works by artificially extending detention time and preserving denitrifying bacteria through the use of substrate added to the mixed liquor and continuously recycled from it prior to secondary clarification. Many other proprietary schemes are being deployed, including DEMON™, Sharon-ANAMMOX™, ANITA-Mox™, and DeAmmon™. The bacteria Brocadia anammoxidans can remove ammonium from waste water through anaerobic oxidation of ammonium to hydrazine, a form of rocket fuel.

Phosphorus Removal

Every adult human excretes between 200 and 1000 grams of phosphorus annually. Studies of United States sewage in the late 1960s estimated mean per capita contributions of 500 grams in urine and feces, 1000 grams in synthetic detergents, and lesser variable amounts used as corrosion and scale control chemicals in water supplies. Source control via alternative detergent formulations has subsequently reduced the largest contribution, but the content of urine and feces will remain unchanged. Phosphorus removal is important as it is a limiting nutrient for algae growth in many fresh water systems. It is also particularly important for water reuse systems where high phosphorus concentrations may lead to fouling of downstream equipment such as reverse osmosis.

Phosphorus can be removed biologically in a process called enhanced biological phosphorus removal. In this process, specific bacteria, called polyphosphate-accumulating organisms (PAOs), are selectively enriched and accumulate large quantities of phosphorus within their cells (up to 20 percent of their mass). When the biomass enriched in these bacteria is separated from the treated water, these biosolids have a high fertilizer value.

Phosphorus removal can also be achieved by chemical precipitation, usually with salts of iron (e.g. ferric chloride), aluminum (e.g. alum), or lime. This may lead to excessive sludge production as hydroxides precipitates and the added chemicals can be expensive. Chemical phosphorus removal requires significantly smaller equipment footprint than biological removal, is easier to operate and is often more reliable than biological phosphorus removal. Another method for phosphorus removal is to use granular laterite.

Once removed, phosphorus, in the form of a phosphate-rich sewage sludge, may be dumped in a landfill or used as fertilizer. In the latter case, the treated sewage sludge is also sometimes referred to as biosolids.

Disinfection

The purpose of disinfection in the treatment of waste water is to substantially reduce the number of microorganisms in the water to be discharged back into the environment for the later use of drinking, bathing, irrigation, etc. The effectiveness of disinfection depends on the quality of the water being treated (e.g., cloudiness, pH, etc.), the type of disinfection being used, the disinfectant dosage (concentration and time), and other environmental variables. Cloudy water will be treated less successfully, since solid matter can shield organisms, especially from ultraviolet light or if contact times are low. Generally, short contact times, low doses and high flows all militate against effective disinfection. Common methods of disinfection include ozone, chlorine, ultraviolet light, or sodium hypochlorite. Chloramine, which is used for drinking water, is not used in the treatment of waste water because of its persistence. After multiple steps of disinfection, the treated water is ready to be released back into the water cycle by means of the nearest body of water or agriculture. Afterwards, the water can be transferred to reserves for everyday human uses.

Chlorination remains the most common form of waste water disinfection in North America due to its low cost and long-term history of effectiveness. One disadvantage is that chlorination of residual organic material can generate chlorinated-organic compounds that may be carcinogenic or harmful to the environment. Residual chlorine or chloramines may also be capable of chlorinating organic material in the natural aquatic environment. Further, because residual chlorine is toxic to aquatic species, the treated effluent must also be chemically dechlorinated, adding to the complexity and cost of treatment.

Ultraviolet (UV) light can be used instead of chlorine, iodine, or other chemicals. Because no chemicals are used, the treated water has no adverse effect on organisms that later consume it, as may be the case with other methods. UV radiation causes damage to the genetic structure of bacteria, viruses, and other pathogens, making them incapable of reproduction. The key disadvantages of UV disinfection are the need for frequent lamp maintenance and replacement and the need for a highly treated effluent to ensure that the target microorganisms are not shielded from the UV radiation (i.e., any solids present in the treated effluent may protect microorganisms from the UV light). In the United Kingdom, UV light is becoming the most common means of disinfection because of the concerns about the impacts of chlorine in chlorinating residual organics in the wastewater and in chlorinating organics in the receiving water. Some sewage treatment systems in Canada and the US also use UV light for their effluent water disinfection.

Ozone (O_3) is generated by passing oxygen (O_2) through a high voltage potential resulting in a third oxygen atom becoming attached and forming O_3. Ozone is very unstable and reactive and oxidizes most organic material it comes in contact with, thereby destroying many pathogenic microorganisms. Ozone is considered to be safer than chlorine because, unlike chlorine which has to be stored on site (highly poisonous in the event of an accidental release), ozone is generated on-site as needed. Ozonation also produces fewer disinfection by-products than chlorination. A disadvantage of ozone disinfection is the high cost of the ozone generation equipment and the requirements for special operators.

Fourth Treatment Stage

Micropollutants such as pharmaceuticals, ingredients of household chemicals, chemicals used in small businesses or industries, environmental persistent pharmaceutical pollutant (EPPP) or pesticides may not be eliminated in the conventional treatment process (primary, secondary and tertiary treatment) and therefore lead to water pollution. Although concentrations of those substances and their decompostion products are quite low, there is still a chance to harm aquatic organisms. For pharmaceuticals, the following substances have been identified as "toxicologically relevant": substances with endocrine disrupting effects, genotoxic substances and substances that enhance the development of bacterial resistances. They mainly belong to the group of environmental persistent pharmaceutical pollutants. Techniques for elimination of micropollutants via a fourth treatment stage during sewage treatment are being tested in Germany, Switzerland and the Netherlands. However, since those techniques are still costly, they are not yet applied on a regular basis. Such process steps mainly consist of activated carbon filters that adsorb the micropollutants. Ozone can also be applied as an oxidative method. Also the use of enzymes such as the enzyme laccase is under investigation. A new concept which could provide an energy-efficient treatment of micropollutants could be the use of laccase secreting fungi cultivated at a wastewater treatment plant to degrade micropollutants and at the same time to provide enzymes at a cathode of a microbial biofuel cells. Microbial biofuel cells are investigated for their property to treat organic matter in wastewater.

To reduce pharmaceuticals in water bodies, also "source control" measures are under investigation, such as innovations in drug development or more responsible handling of drugs.

Odor Control

Odors emitted by sewage treatment are typically an indication of an anaerobic or "septic" condition. Early stages of processing will tend to produce foul-smelling gases, with hydrogen sulfide being most common in generating complaints. Large process plants in urban areas will often treat the odors with carbon reactors, a contact media with bio-slimes, small doses of chlorine, or circulating fluids to biologically capture and metabolize the noxious gases. Other methods of odor control exist, including addition of iron salts, hydrogen peroxide, calcium nitrate, etc. to manage hydrogen sulfide levels.

High-density solids pumps are suitable for reducing odors by conveying sludge through hermetic closed pipework.

Energy Requirements

For conventional sewage treatment plants, around 30 percent of the annual operating costs is usually required for energy. The energy requirements vary with type of treatment process as well as wastewater load. For example, constructed wetlands have a lower energy requirement than activated sludge plants, as less energy is required for the aeration step. Sewage treatment plants that produce biogas in their sewage sludge treatment process with anaerobic digestion can produce enough energy to meet most of the energy needs of the sewage treatment plant itself.

In conventional secondary treatment processes, most of the electricity is used for aeration, pumping systems and equipment for the dewatering and drying of sewage sludge. Advanced wastewater treatment plants, e.g. for nutrient removal, require more energy than plants that only achieve primary or secondary treatment.

Sludge Treatment and Disposal

The sludges accumulated in a wastewater treatment process must be treated and disposed of in a safe and effective manner. The purpose of digestion is to reduce the amount of organic matter and the number of disease-causing microorganisms present in the solids. The most common treatment options include anaerobic digestion, aerobic digestion, and composting. Incineration is also used, albeit to a much lesser degree. Sludge treatment depends on the amount of solids generated and other site-specific conditions. Composting is most often applied to small-scale plants with aerobic digestion for mid-sized operations, and anaerobic digestion for the larger-scale operations.

The sludge is sometimes passed through a so-called pre-thickener which de-waters the sludge. Types of pre-thickeners include centrifugal sludge thickeners rotary drum sludge thickeners and belt filter presses. Dewatered sludge may be incinerated or transported offsite for disposal in a landfill or use as an agricultural soil amendment.

Environment Aspects

Many processes in a wastewater treatment plant are designed to mimic the natural treatment processes that occur in the environment, whether that environment is a natural water body or the ground. If not overloaded, bacteria in the environment will consume organic contaminants, although this will reduce the levels of oxygen in the water and may significantly change the overall ecology of the receiving water. Native bacterial populations feed on the organic contaminants, and the numbers of disease-causing microorganisms are reduced by natural environmental conditions such as predation or exposure to ultraviolet radiation. Consequently, in cases where the receiving environment provides a high level of dilution, a high degree of wastewater treatment may not be required. However, recent evidence has demonstrated that very low levels of specific contaminants in wastewater, including hormones (from animal husbandry and residue from human hormonal contraception methods) and synthetic materials such as phthalates that mimic hormones in their action, can have an unpredictable adverse impact on the natural biota and potentially on humans if the water is re-used for drinking water. In the US and EU, uncontrolled discharges of wastewater to the environment are not permitted under law, and strict water quality requirements are to be met, as clean drinking water is essential. A significant threat in the coming decades will be the increasing uncontrolled discharges of wastewater within rapidly developing countries.

The outlet of the Karlsruhe sewage treatment plant flows into the Alb.

Effects on Biology

Sewage treatment plants can have multiple effects on nutrient levels in the water that the treated sewage flows into. These nutrients can have large effects on the biological life in the water in contact with the effluent. Stabilization ponds (or sewage treatment ponds) can include any of the following:

- Oxidation ponds, which are aerobic bodies of water usually 1–2 meters in depth that receive effluent from sedimentation tanks or other forms of primary treatment.

  - Dominated by algae

- Polishing ponds are similar to oxidation ponds but receive effluent from an oxidation pond or from a plant with an extended mechanical treatment.

  - Dominated by zooplankton

- Facultative lagoons, raw sewage lagoons, or sewage lagoons are ponds where sewage is added with no primary treatment other than coarse screening. These ponds provide effective treatment when the surface remains aerobic; although anaerobic conditions may develop near the layer of settled sludge on the bottom of the pond.

- Anaerobic lagoons are heavily loaded ponds.

  - Dominated by bacteria

- Sludge lagoons are aerobic ponds, usually 2 to 5 meters in depth, that receive anaerobically digested primary sludge, or activated secondary sludge under water.

  - Upper layers are dominated by algae

Phosphorus limitation is a possible result from sewage treatment and results in flagellate-dominated plankton, particularly in summer and fall.

A phytoplankton study found high nutrient concentrations linked to sewage effluents. High nutrient concentration leads to high chlorophyll a concentrations, which is a proxy for primary production in marine environments. High primary production means high phytoplankton populations and most likely high zooplankton populations, because zooplankton feed on phytoplankton. However, effluent released into marine systems also leads to greater population instability.

The planktonic trends of high populations close to input of treated sewage is contrasted by the bacterial trend. In a study of *Aeromonas* spp. in increasing distance from a wastewater source, greater change in seasonal cycles was found the furthest from the effluent. This trend is so strong that the furthest location studied actually had an inversion of the *Aeromonas* spp. cycle in comparison to that of fecal coliforms. Since there is a main pattern in the cycles that occurred simultaneously at all stations it indicates seasonal factors (temperature, solar radiation, phytoplankton) control of the bacterial population. The effluent dominant species changes from *Aeromonas caviae* in winter to *Aeromonas sobria* in the spring and fall while the inflow dominant species is *Aeromonas caviae*, which is constant throughout the seasons.

Treated Sewage Reuse

With suitable technology, it is possible to reuse sewage effluent for drinking water, although this is usually only done in places with limited water supplies, such as Windhoek and Singapore.

In Israel, about 50 percent of agricultural water use (total use was 1 billion cubic metres in 2008) is provided through reclaimed sewer water. Future plans call for increased use of treated sewer water as well as more desalination plants.

Sewage Treatment in Developing Countries

Few reliable figures exist on the share of the wastewater collected in sewers that is being treated in the world. A global estimate by UNDP and UN-Habitat is that 90% of all wastewater generated is released into the environment untreated. In many developing countries the bulk of domestic and industrial wastewater is discharged without any treatment or after primary treatment only.

In Latin America about 15 percent of collected wastewater passes through treatment plants (with varying levels of actual treatment). In Venezuela, a below average country in South America with respect to wastewater treatment, 97 percent of the country's sewage is discharged raw into the environment. In Iran, a relatively developed Middle Eastern country, the majority of Tehran's population has totally untreated sewage injected to the city's groundwater. However, the construction of major parts of the sewage system, collection and treatment, in Tehran is almost complete, and under development, due to be fully completed by the end of 2012. In Isfahan, Iran's third largest city, sewage treatment was started more than 100 years ago.

Only few cities in sub-Saharan Africa have sewer-based sanitation systems, let alone wastewater treatment plants, an exception being South Africa and – until the late 1990s- Zimbabwe. Instead, most urban residents in sub-Saharan Africa rely on onsite sanitation systems without sewers, such as septic tanks and pit latrines, and faecal sludge management in these cities is an enormous challenge.

History

Basic sewer systems were used for waste removal in ancient Mesopotamia, where vertical shafts carried the waste away into cesspools. Similar systems existed in the Indus Valley civilization in modern-day India and in Ancient Crete and Greece. In the Middle Ages the sewer systems built by the Romans fell into disuse and waste was collected into cesspools that were periodically emptied by workers known as 'rakers' who would often sell it as fertilizer to farmers outside the city.\

Modern sewage systems were first built in the mid-nineteenth century as a reaction to the exacerbation of sanitary conditions brought on by heavy industrialization and urbanization. Due to the contaminated water supply, cholera outbreaks occurred in 1832, 1849 and 1855 in London, killing tens of thousands of people. This, combined with the Great Stink of 1858, when the smell of untreated human waste in the River Thames became overpowering, and the report into sanitation reform of the Royal Commissioner Edwin Chadwick, led to the Metropolitan Commission of Sewers appointing Sir Joseph Bazalgette to construct a vast underground sewage system for the safe removal of waste. Contrary to Chadwick's recommendations, Bazalgette's system, and others later built in Continental Europe, did not pump the sewage onto farm land for use as fertilizer; it was simply piped to a natural waterway away from population centres, and pumped back into the environment.

FARADAY GIVING HIS CARD TO FATHER THAMES;
And we hope the Dirty Fellow will consult the learned Professor.

The Great Stink of 1858 stimulated research into the problem of sewage treatment. In this caricature in *The Times*, Michael Faraday reports to *Father Thames* on the state of the river.

Early Attempts

One of the first attempts at diverting sewage for use as a fertilizer in the farm was made by the cotton mill owner James Smith in the 1840s. He experimented with a piped distribution system initially proposed by James Vetch that collected sewage from his factory and pumped it into the outlying farms, and his success was enthusiastically followed by Edwin Chadwick and supported by organic chemist Justus von Liebig.

The idea was officially adopted by the Health of Towns Commission, and various schemes (known as sewage farms) were trialled by different municipalities over the next 50 years. At first, the heavier solids were channeled into ditches on the side of the farm and were covered over when full, but soon flat-bottomed tanks were employed as reservoirs for the sewage; the earliest patent was taken out by William Higgs in 1846 for "tanks or reservoirs in which the contents of sewers and drains from cities, towns and villages are to be collected and the solid animal or vegetable matters therein contained, solidified and dried..." Improvements to the design of the tanks included the introduction of the horizontal-flow tank in the 1850s and the radial-flow tank in 1905. These tanks had to be manually de-sludged periodically, until the introduction of automatic mechanical de-sludgers in the early 1900s.

The precursor to the modern septic tank was the cesspool in which the water was sealed off to prevent contamination and the solid waste was slowly liquified due to anaerobic action; it was invented by L.H Mouras in France in the 1860s. Donald Cameron, as City Surveyor for Exeter patented an improved version in 1895, which he called a 'septic tank'; septic having the meaning of 'bacterial'. These are still in worldwide use, especially in rural areas unconnected to large-scale sewage systems.

Chemical Treatment

It was not until the late 19th century that it became possible to treat the sewage by chemically breaking it down through the use of microorganisms and removing the pollutants. Land treatment was also steadily becoming less feasible, as cities grew and the volume of sewage produced could no longer be absorbed by the farmland on the outskirts.

Sir Edward Frankland, a distinguished chemist, who demonstrated the possibility of chemically treating sewage in the 1870s.

Sir Edward Frankland conducted experiments at the Sewage Farm in Croydon, England, during the 1870s and was able to demonstrate that filtration of sewage through porous gravel produced a nitrified effluent (the ammonia was converted into nitrate) and that the filter remained unclogged over long periods of time. This established the then revolutionary possibility of biological treatment of sewage using a contact bed to oxidize the waste. This concept was taken up by the chief chemist for the London Metropolitan Board of Works, William Libdin, in 1887:

> ...in all probability the true way of purifying sewage...will be first to separate the sludge, and then turn into neutral effluent... retain it for a sufficient period, during which time it should be fully aerated, and finally discharge it into the stream in a purified condition. This is indeed what is aimed at and imperfectly accomplished on a sewage farm.

From 1885 to 1891 filters working on this principle were constructed throughout the UK and the idea was also taken up in the US at the Lawrence Experiment Station in Massachusetts, where Frankland's work was confirmed. In 1890 the LES developed a 'trickling filter' that gave a much more reliable performance.

Contact beds were developed in Salford, Lancashire and by scientists working for the London City Council in the early 1890s. According to Christopher Hamlin, this was part of a conceptual revolution that replaced the philosophy that saw "sewage purification as the prevention of decomposition with one that tried to facilitate the biological process that destroy sewage naturally."

Contact beds were tanks containing the inert substance, such as stones or slate, that maximized the surface area available for the microbial growth to break down the sewage. The sewage was held in the tank until it was fully decomposed and it was then filtered out into the ground. This method quickly became widespread, especially in the UK, where it was used in Leicester, Sheffield, Manchester and Leeds. The bacterial bed was simultaneously developed by Joseph Corbett as Borough Engineer in Salford and experiments in 1905 showed that his method was superior in that greater volumes of sewage could be purified better for longer periods of time than could be achieved by the contact bed.

The Royal Commission on Sewage Disposal published its eighth report in 1912 that set what became the international standard for sewage discharge into rivers; the '20:30 standard', which allowed 20 mg Biochemical oxygen demand and 30 mg suspended solid per litre.

Sedimentation (Water Treatment)

Sedimentation is a physical water treatment process using gravity to remove suspended solids from water. Solid particles entrained by the turbulence of moving water may be removed naturally by sedimentation in the still water of lakes and oceans. Settling basins are ponds constructed for the purpose of removing entrained solids by sedimentation. Clarifiers are tanks built with mechanical means for continuous removal of solids being deposited by sedimentation.

Basics

Suspended solids (or SS), is the mass of dry solids retained by a filter of a given porosity related to the volume of the water sample. This includes particles of a size not lower than 10 μm.

Colloids are particles of a size between 0.001 μm and 1 μm depending on the method of quantification. Because of Brownian motion and electrostatic forces balancing the gravity, they are not likely to settle naturally.

The limit sedimentation velocity of a particle is its theoretical descending speed in clear and still water. In settling process theory, a particle will settle only if:

1. In a vertical ascending flow, the ascending water velocity is lower than the limit sedimentation velocity.

2. In a longitudinal flow, the ratio of the length of the tank to the height of the tank is higher than the ratio of the water velocity to the limit sedimentation velocity.

Removal of suspended particles by sedimentation depends upon the size and specific gravity of those particles. Suspended solids retained on a filter may remain in suspension if their specific gravity is similar to water while very dense particles passing through the filter may settle. Settleable solids are measured as the visible volume accumulated at the bottom of an Imhoff cone after water has settled for one hour.

Gravitational theory is employed, alongside the derivation from Newton's second law and the Navier–Stokes equations.

$$V_s = \sqrt{4/3\ ((\rho_p - \rho_d)/\rho_p\ )}\ (gd_p)/C_d\ (1)$$

Stokes' law explains the relationship between the settling rate and the particle diameter. Under specific conditions, the particle settling rate is directly proportional to the square of particle diameter and inversely proportional to liquid viscosity.

The settling velocity, defined as the residence time taken for the particles to settle in the tank, enables the calculation of tank volume. Precise design and operation of a sedimentation tank is of high importance in order to keep the amount of sediment entering the diversion system to a minimum threshold by maintaining the transport system and stream stability to remove the sediment diverted from the system. This is achieved by reducing stream velocity as low as possible for the longest period of time possible. This is feasible by widening the approach channel and lowering its floor to reduce flow velocity thus allowing sediment to settle out of suspension due to gravity. The settling behavior of heavier particulates is also affected by the turbulence.

Designs

Although sedimentation might occur in tanks of other shapes, removal of accumulated solids is easiest with conveyor belts in rectangular tanks or with scrapers rotating around the central axis of circular tanks. Settling basins and clarifiers should be designed based on the settling velocity of the smallest particle to be theoretically 100% removed. The overflow rate is defined as:

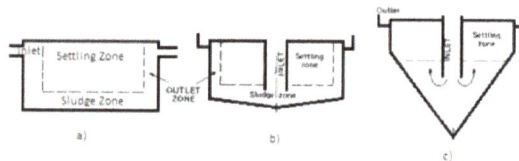

Figure 1. Different clarifier designs

Overflow rate (V_o) = Flow of water (Q (cubic metre per second)) /(Surface area of settling basin (A))(m^2)

The unit of overflow rate is usually feet per second, a velocity. Any particle with settling velocity (Vs) greater than the overflow rate will settle out, while other particles will settle in the ratio Vs/Vo. There are recommendations on the overflow rates for each design that ideally take into account the change in particle size as the solids move through the operation:

- Quiescent zones: 0.031 ft/s

- Full-flow basins: 0.013 ft/s

- Off-line basins: 0.0015 ft/s

However, factors such as flow surges, wind shear, scour, and turbulence reduce the effectiveness of settling. To compensate for these less than ideal conditions, it is recommended doubling the area calculated by the previous equation. It is also important to equalize flow distribution at each point across the cross-section of the basin. Poor inlet and outlet designs can produce extremely poor flow characteristics for sedimentation.

Settling basins and clarifiers can be designed as long rectangles (Figure 1.a), that are hydraulically more stable and easier to control for large volumes. Circular clarifiers (Fig. 1.b) work as a common thickener (without the usage of rakes), or as upflow tanks (Fig. 1.c).

Sedimentation efficiency does not depend on the tank depth. If the forward velocity is low enough so that the settled material does not re-suspend from the tank floor, the area is still the main parameter when designing a settling basin or clarifier, taking care that the depth is not too low.

Assessment of Main Process Characteristics

Settling basins and clarifiers are designed to retain water so that suspended solids can settle. By sedimentation principles, the suitable treatment technologies should be chosen depending on the specific gravity, size and shear resistance of particles. Depending on the size and density of particles, and physical properties of the solids, there are four types of sedimentation processes:

- Type 1 – Dilutes, non-flocculent, free-settling (every particle settles independently.)

- Type 2 – Dilute, flocculent (particles can flocculate as they settle).

- Type 3 – Concentrated suspensions, zone settling, hindered settling (sludge thickening).

- Type 4 – Concentrated suspensions, compression (sludge thickening).

Different factors control the sedimentation rate in each.

Settling of Discrete Particles

Unhindered settling is a process that removes the discrete particles in a very low concentration without interference from nearby particles. In general, if the concentration
of the solutions is lower than 500 mg/L total suspended solids, sedimentation will be
considered discrete. Concentrations of raceway effluent total suspended solids (TSS) in
the west are usually less than 5 mg/L net. TSS concentrations of off-line settling basin
effluent are less than 100 mg/L net. The particles keep their size and shape during discrete settling, with an independent velocity. With such low concentrations of suspended particles, the probability of particle collisions is very low and consequently the rate
of flocculation is small enough to be neglected for most calculations. Thus the surface
area of the settling basin becomes the main factor of sedimentation rate. All continuous
flow settling basins are divided into four parts: inlet zone, settling zone, sludge zone
and outlet zone (Figure 2).

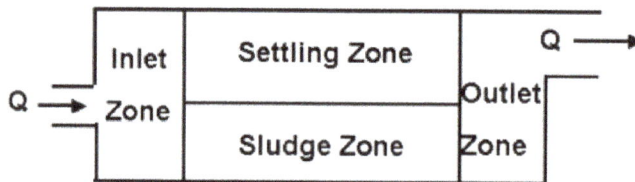

Figure 2. The four functional zones of a continuous flow settling basin

In the inlet zone, flow is established in a same forward direction. Sedimentation
occurs in the settling zone as the water flow towards to outlet zone. The clarified
liquid is then flow out from outlet zone. Sludge zone: settled will be collected here
and usually we assume that it is removed from water flow once the particles arrives
the sludge zone.

In an ideal rectangular sedimentation tank, in the settling zone, the critical particle
enters at the top of the settling zone, and the settle velocity would be the smallest value
to reach the sludge zone, and at the end of outlet zone, the velocity component of this
critical particle are Vs, the settling velocity in vertical direction and Vh in horizontal
direction.

From Figure 1, the time needed for the particle to settle;

$$t_o = H/V_p = L/V_s \quad (3)$$

Since the surface area of the tank is WL, and $V_s = Q/WL$, $V_h = Q/WH$, where Q is the
flow rate and W, L, H is the width, length, depth of the tank.

According to Eq. 1, this also is a basic factor that can control the sedimentation tank
performance which called overflow rate.

Eq. 2 also tell us that the depth of sedimentation tank is independent to the sedimentation efficiency, only if the forward velocity is low enough to make sure the settled mass would not suspended again from the tank floor.

Settlement of Flocculent Particles

In a horizontal sedimentation tank, some particles may not follow the diagonal line in Fig. 1, while settling faster as they grow. So this says that particles can grow and develop a higher settling velocity if a greater depth with longer retention time. However, the collision chance would be even greater if the same retention time were spread over a longer, shallower tank. In fact, in order to avoid hydraulic short-circuiting, tanks usually are made 3–6 m deep with retention times of a few hours.

Zone-settling Behaviour

As the concentration of particles in a suspension is increased, a point is reached where particles are so close together that they no longer settle independently of one another and the velocity fields of the fluid displaced by adjacent particles, overlap. There is also a net upward flow of liquid displaced by the settling particles. This results in a reduced particle-settling velocity and the effect is known as hindered settling.

There is a common case for hindered settling occurs. the whole suspension tends to settle as a 'blanket' due to its extremely high particle concentration. This is known as zone settling, because it is easy to make a distinction between several different zones which separated by concentration discontinuities. Fig. 3 represents a typical batch-settling column tests on a suspension exhibiting zone-settling characteristics. There is a clear interface near the top of the column would be formed to separating the settling sludge mass from the clarified supernatant as long as leaving such a suspension to stand in a settling column. As the suspension settles, this interface will move down at the same speed. At the same time, there is an interface near the bottom between that settled suspension and the suspended blanket. After settling of suspension is complete, the bottom interface would move upwards and meet the top interface which moves downwards.

Compression Settling

The settling particles can contact each other and arise when approaching the floor of the sedimentation tanks at very high particle concentration. So that further settling will only occur in adjust matrix as the sedimentation rate decreasing. This is can be illustrated by the lower region of the zone-settling diagram (Figure 3). In Compression zone, the settled solids are compressed by gravity (the weight of solids), as the settled solids are compressed under the weight of overlying solids, and water is squeezed out while the space gets smaller.

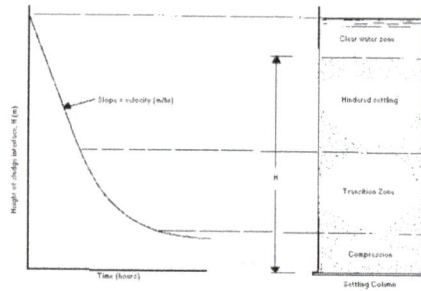

Figure3: Typical batch-settling column test on a suspension exhibiting zone-settling characteristics

Applications

Potable Water Treatment

Sedimentation in potable water treatment generally follows a step of chemical coagulation and flocculation, which allows grouping particles together into flocs of a bigger size. This increases the settling speed of suspended solids and allows settling colloids.

Wastewater Treatment

Sedimentation has been used to treat wastewater for millennia.

Primary treatment of sewage is removal of floating and settleable solids through sedimentation. *Primary clarifiers* reduce the content of suspended solids as well as the pollutant embedded in the suspended solids. Because of the large amount of reagent necessary to treat domestic wastewater, preliminary chemical coagulation and flocculation are generally not used, remaining suspended solids being reduced by following stages of the system. However, coagulation and flocculation can be used for building a compact treatment plant (also called a "package treatment plant"), or for further polishing of the treated water.

Sedimentation tanks called "secondary clarifiers" remove flocs of biological growth created in some methods of secondary treatment including activated sludge, trickling filters and rotating biological contactors.

Industrial Wastewater Treatment

Industrial wastewater treatment covers the mechanisms and processes used to treat wastewater that is produced as a by-product of industrial or commercial activities. After treatment, the treated industrial wastewater (or effluent) may be reused or released to a sanitary sewer or to a surface water in the environment. Most industries produce some wastewater although recent trends in the developed world have been to minimise such production or recycle such wastewater within the production process. However, many industries remain dependent on processes that produce wastewaters.

Sources of Industrial Wastewater

Complex Organic Chemicals Industry

A range of industries manufacture or use complex organic chemicals. These include pesticides, pharmaceuticals, paints and dyes, petrochemicals, detergents, plastics, paper pollution, etc. Waste waters can be contaminated by feedstock materials, by-products, product material in soluble or particulate form, washing and cleaning agents, solvents and added value products such as plasticisers. Treatment facilities that do not need control of their effluent typically opt for a type of aerobic treatment, i.e. aerated lagoons.

Electric Power Plants

Fossil-fuel power stations, particularly coal-fired plants, are a major source of industrial wastewater. Many of these plants discharge wastewater with significant levels of metals such as lead, mercury, cadmium and chromium, as well as arsenic, selenium and nitrogen compounds (nitrates and nitrites). Wastewater streams include flue-gas desulfurization, fly ash, bottom ash and flue gas mercury control. Plants with air pollution controls such as wet scrubbers typically transfer the captured pollutants to the wastewater stream.

Ash ponds, a type of surface impoundment, are a widely used treatment technology at coal-fired plants. These ponds use gravity to settle out large particulates (measured as total suspended solids) from power plant wastewater. This technology does not treat dissolved pollutants. Power stations use additional technologies to control pollutants, depending on the particular wastestream in the plant. These include dry ash handling, closed-loop ash recycling, chemical precipitation, biological treatment (such as an activated sludge process), and evaporation.

Food Industry

Wastewater generated from agricultural and food operations has distinctive characteristics that set it apart from common municipal wastewater managed by public or private sewage treatment plants throughout the world: it is biodegradable and non-toxic, but has high concentrations of biochemical oxygen demand (BOD) and suspended solids (SS). The constituents of food and agriculture wastewater are often complex to predict, due to the differences in BOD and pH in effluents from vegetable, fruit, and meat products and due to the seasonal nature of food processing and post-harvesting.

Processing of food from raw materials requires large volumes of high grade water. Vegetable washing generates waters with high loads of particulate matter and some dissolved organic matter. It may also contain surfactants.

Animal slaughter and processing produces very strong organic waste from body fluids, such as blood, and gut contents. This wastewater is frequently contaminated by significant levels of antibiotics and growth hormones from the animals and by a variety of pesticides used to control external parasites.

Processing food for sale produces wastes generated from cooking which are often rich in plant organic material and may also contain salt, flavourings, colouring material and acids or alkali. Very significant quantities of oil or fats may also be present.

Iron and Steel Industry

The production of iron from its ores involves powerful reduction reactions in blast furnaces. Cooling waters are inevitably contaminated with products especially ammonia and cyanide. Production of coke from coal in coking plants also requires water cooling and the use of water in by-products separation. Contamination of waste streams includes gasification products such as benzene, naphthalene, anthracene, cyanide, ammonia, phenols, cresols together with a range of more complex organic compounds known collectively as polycyclic aromatic hydrocarbons (PAH).

The conversion of iron or steel into sheet, wire or rods requires hot and cold mechanical transformation stages frequently employing water as a lubricant and coolant. Contaminants include hydraulic oils, tallow and particulate solids. Final treatment of iron and steel products before onward sale into manufacturing includes *pickling* in strong mineral acid to remove rust and prepare the surface for tin or chromium plating or for other surface treatments such as galvanisation or painting. The two acids commonly used are hydrochloric acid and sulfuric acid. Wastewaters include acidic rinse waters together with waste acid. Although many plants operate acid recovery plants (particularly those using hydrochloric acid), where the mineral acid is boiled away from the iron salts, there remains a large volume of highly acid ferrous sulfate or ferrous chloride to be disposed of. Many steel industry wastewaters are contaminated by hydraulic oil, also known as *soluble oil*.

Mines and Quarries

The principal waste-waters associated with mines and quarries are slurries of rock particles in water. These arise from rainfall washing exposed surfaces and haul roads and also from rock washing and grading processes. Volumes of water can be very high, especially rainfall related arisings on large sites. Some specialized separation operations, such as coal washing to separate coal from native rock using density gradients, can produce wastewater contaminated by fine particulate haematite and surfactants. Oils and hydraulic oils are also common contaminants.

Wastewater from metal mines and ore recovery plants are inevitably contaminated by the minerals present in the native rock formations. Following crushing and extraction of the desirable materials, undesirable materials may enter the wastewater stream. For metal mines, this can include unwanted metals such as zinc and other materials such as arsenic. Extraction of high value metals such as gold and silver may generate slimes containing very fine particles in where physical removal of contaminants becomes particularly difficult.

Mine wastewater effluent in Peru, with neutralized pH from tailing runoff.

Additionally, the geologic formations that harbour economically valuable metals such as copper and gold very often consist of sulphide-type ores. The processing entails grinding the rock into fine particles and then extracting the desired metal(s), with the leftover rock being known as tailings. These tailings contain a combination of not only undesirable leftover metals, but also sulphide components which eventually form sulphuric acid upon the exposure to air and water that inevitably occurs when the tailings are disposed of in large impoundments. The resulting acid mine drainage, which is often rich in heavy metals (because acids dissolve metals), is one of the many environmental impacts of mining.

Nuclear Industry

The waste production from the nuclear and radio-chemicals industry is dealt with as *Radioactive waste.*

Pulp and Paper Industry

Effluent from the pulp and paper industry is generally high in suspended solids and BOD. Plants that bleach wood pulp for paper making may generate chloroform, dioxins (including 2,3,7,8-TCDD), furans, phenols and chemical oxygen demand (COD). Stand-alone paper mills using imported pulp may only require simple primary treatment, such as sedimentation or dissolved air flotation. Increased BOD or COD loadings, as well as organic pollutants, may require biological treatment such as activated sludge or upflow anaerobic sludge blanket reactors. For mills with high inorganic loadings like salt, tertiary treatments may be required, either general membrane treatments like ultrafiltration or reverse osmosis or treatments to remove specific contaminants, such as nutrients.

Industrial Oil Contamination

Industrial applications where oil enters the wastewater stream may include vehicle wash bays, workshops, fuel storage depots, transport hubs and power generation. Often the wastewater is discharged into local sewer or trade waste systems and must meet local environmental specifications. Typical contaminants can include solvents, detergents, grit. lubricants and hydrocarbons.

Water Treatment

Many industries have a need to treat water to obtain very high quality water for demanding purposes such as environmental discharge compliance. Water treatment produces organic and mineral sludges from filtration and sedimentation. Ion exchange using natural or synthetic resins removes calcium, magnesium and carbonate ions from water, typically replacing them with sodium, chloride, hydroxyl and/or other ions. Regeneration of ion exchange columns with strong acids and alkalis produces a wastewater rich in hardness ions which are readily precipitated out, especially when in admixture with other wastewater constituents.

Wool Processing

Insecticide residues in fleeces are a particular problem in treating waters generated in wool processing. Animal fats may be present in the wastewater, which if not contaminated, can be recovered for the production of tallow or further rendering.

Treatment of Industrial Wastewater

The various types of contamination of wastewater require a variety of strategies to remove the contamination.

Brine Treatment

Brine treatment involves removing dissolved salt ions from the waste stream. Although similarities to seawater or brackish water desalination exist, industrial brine treatment may contain unique combinations of dissolved ions, such as hardness ions or other metals, necessitating specific processes and equipment.

Brine treatment systems are typically optimized to either reduce the volume of the final discharge for more economic disposal (as disposal costs are often based on volume) or maximize the recovery of fresh water or salts. Brine treatment systems may also be optimized to reduce electricity consumption, chemical usage, or physical footprint.

Brine treatment is commonly encountered when treating cooling tower blowdown, produced water from steam assisted gravity drainage (SAGD), produced water from natural gas extraction such as coal seam gas, frac flowback water, acid mine or acid rock drainage, reverse osmosis reject, chlor-alkali wastewater, pulp and paper mill effluent, and waste streams from food and beverage processing.

Brine treatment technologies may include: membrane filtration processes, such as reverse osmosis; ion exchange processes such as electrodialysis or weak acid cation exchange; or evaporation processes, such as brine concentrators and crystallizers employing mechanical vapour recompression and steam.

Reverse osmosis may not be viable for brine treatment, due to the potential for fouling caused by hardness salts or organic contaminants, or damage to the reverse osmosis membranes from hydrocarbons.

Evaporation processes are the most widespread for brine treatment as they enable the highest degree of concentration, as high as solid salt. They also produce the highest purity effluent, even distillate-quality. Evaporation processes are also more tolerant of organics, hydrocarbons, or hardness salts. However, energy consumption is high and corrosion may be an issue as the prime mover is concentrated salt water. As a result, evaporation systems typically employ titanium or duplex stainless steel materials.

Brine Management

Brine management examines the broader context of brine treatment and may include consideration of government policy and regulations, corporate sustainability, environmental impact, recycling, handling and transport, containment, centralized compared to on-site treatment, avoidance and reduction, technologies, and economics. Brine management shares some issues with leachate management and more general waste management.

Solids Removal

Most solids can be removed using simple sedimentation techniques with the solids recovered as slurry or sludge. Very fine solids and solids with densities close to the density of water pose special problems. In such case filtration or ultrafiltration may be required. Although, flocculation may be used, using alum salts or the addition of polyelectrolytes.

Oils and Grease Removal

The effective removal of oils and grease is dependent on the characteristics of the oil in terms of its suspension state and droplet size, which will in turn affect the choice of separator technology.

Oil pollution in water usually comes in four states, often in combination:

- free oil - large oil droplets sitting on the surface;

- heavy oil, which sits at the bottom, often adhering to solids like dirt;

- emulsified, where the oil droplets are heavily "chopped"; and

- dissolved oil, where the droplets are fully dispersed and not visible. Emulsified oil droplets are the most common in industrial oily wastewater and are extremely difficult to separate.

The methodology for separating the oil is dependent on the oil droplet size. Larger oil droplets such as those in free oil pollution are easily removed, but as the droplets become smaller, some separator technologies perform better than others.

Most separator technologies will have an optimum range of oil droplet sizes that can be effectively treated. This is known as the "micron rating."

Analysing the oily water to determine droplet size can be performed with a video particle analyser. Alternatively, there are commonalities in industries for oil droplet sizes. Larger droplets–greater than 60 microns–are often present in wastewater in workshops, re-fuel areas and depots. Twenty to 50 micron oil droplets often are present in vehicle wash bays, meat processing and dairy manufacturing effluent and aluminium billet cooling towers. Smaller droplets in the range of 10 to 20 microns tend to occur in workshops and condensates.

Each separator technology will have its' own performance curve outlining optimum performance based on oil droplet size. the most common separators are gravity tanks or pits, API oil-water separators or plate packs, chemical treatment via DAFs, centrifuges, media filters and hydrocyclones.

API Separators

Many oils can be recovered from open water surfaces by skimming devices. Considered a dependable and cheap way to remove oil, grease and other hydrocarbons from water, oil skimmers can sometimes achieve the desired level of water purity. At other times, skimming is also a cost-efficient method to remove most of the oil before using membrane filters and chemical processes. Skimmers will prevent filters from blinding prematurely and keep chemical costs down because there is less oil to process.

Because grease skimming involves higher viscosity hydrocarbons, skimmers must be equipped with heaters powerful enough to keep grease fluid for discharge. If floating grease forms into solid clumps or mats, a spray bar, aerator or mechanical apparatus can be used to facilitate removal.

However, hydraulic oils and the majority of oils that have degraded to any extent will also have a soluble or emulsified component that will require further treatment to eliminate. Dissolving or emulsifying oil using surfactants or solvents usually exacerbates the problem rather than solving it, producing wastewater that is more difficult to treat.

The wastewaters from large-scale industries such as oil refineries, petrochemical plants, chemical plants, and natural gas processing plants commonly contain gross amounts of oil and suspended solids. Those industries use a device known as an API oil-water separator which is designed to separate the oil and suspended solids from their wastewater effluents. The name is derived from the fact that such separators are designed according to standards published by the American Petroleum Institute (API).

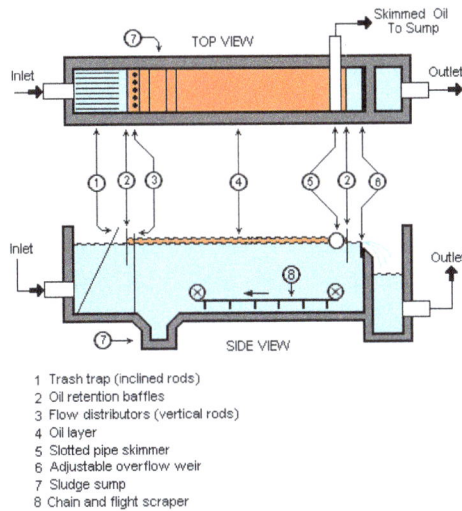

1 Trash trap (inclined rods)
2 Oil retention baffles
3 Flow distributors (vertical rods)
4 Oil layer
5 Slotted pipe skimmer
6 Adjustable overflow weir
7 Sludge sump
8 Chain and flight scraper

A typical API oil-water separator used in many industries

The API separator is a gravity separation device designed by using Stokes Law to define the rise velocity of oil droplets based on their density and size. The design is based on the specific gravity difference between the oil and the wastewater because that difference is much smaller than the specific gravity difference between the suspended solids and water. The suspended solids settles to the bottom of the separator as a sediment layer, the oil rises to top of the separator and the cleansed wastewater is the middle layer between the oil layer and the solids.

Typically, the oil layer is skimmed off and subsequently re-processed or disposed of, and the bottom sediment layer is removed by a chain and flight scraper (or similar device) and a sludge pump. The water layer is sent to further treatment for additional removal of any residual oil and then to some type of biological treatment unit for removal of undesirable dissolved chemical compounds.

Parallel plate separators are similar to API separators but they include tilted parallel plate assemblies (also known as parallel packs). The parallel plates provide more surface for suspended oil droplets to coalesce into larger globules. Such separators still depend upon the specific gravity between the suspended oil and the water. However, the parallel plates enhance the degree of oil-water separation. The result is that a parallel plate separator requires significantly less space than a conventional API separator to achieve the same degree of separation.

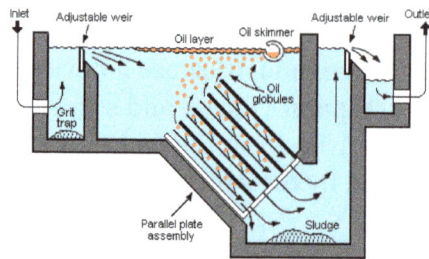

A typical parallel plate separator

Hydrocyclone Oil Separators

Hydrocyclone oil separators operate on the process where wastewater enters the cyclone chamber and is spun under extreme centrifugal forces more than 1000 times the force of gravity. This force causes the water and oil droplets to separate. The separated oil is discharged from one end of the cyclone where treated water is discharged through the opposite end for further treatment, filtration or discharge.

Hydrocyclones are useful for the greatest range of oil droplet sizes operating from less than 10 microns and up and can operate continuously without water pre-treatment and at any temperature and pH. Applications where hydrocyclones are found are in industry where oily water sources arise in workshops, vehicle wash bays, transport hubs, fuel depots and aluminium billet processing. Animal fats from meat processing and dairy manufacturing can also be removed without the need of chemical treatment that often is required for dissolved air flotation (DAF) systems.

Removal of Biodegradable Organics

Biodegradable organic material of plant or animal origin is usually possible to treat using extended conventional sewage treatment processes such as activated sludge or trickling filter. Problems can arise if the wastewater is excessively diluted with washing water or is highly concentrated such as undiluted blood or milk. The presence of cleaning agents, disinfectants, pesticides, or antibiotics can have detrimental impacts on treatment processes.

Activated Sludge Process

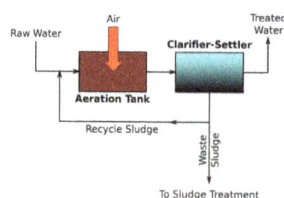

A generalized diagram of an activated sludge process.

Activated sludge is a biochemical process for treating sewage and industrial wastewater that uses air (or oxygen) and microorganisms to biologically oxidize organic pollutants, producing a waste sludge (or floc) containing the oxidized material. In general, an activated sludge process includes:

- An aeration tank where air (or oxygen) is injected and thoroughly mixed into the wastewater.

- A settling tank (usually referred to as a clarifier or "settler") to allow the waste sludge to settle. Part of the waste sludge is recycled to the aeration tank and the remaining waste sludge is removed for further treatment and ultimate disposal.

Trickling Filter Process

A trickling filter consists of a bed of rocks, gravel, slag, peat moss, or plastic media over which wastewater flows downward and contacts a layer (or film) of microbial slime covering the bed media. Aerobic conditions are maintained by forced air flowing through the bed or by natural convection of air. The process involves adsorption of organic compounds in the wastewater by the microbial slime layer, diffusion of air into the slime layer to provide the oxygen required for the biochemical oxidation of the organic compounds. The end products include carbon dioxide gas, water and other products of the oxidation. As the slime layer thickens, it becomes difficult for the air to penetrate the layer and an inner anaerobic layer is formed.

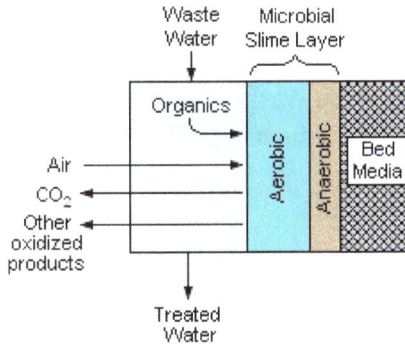

Image 1: A schematic cross-section of the contact face of the bed media in a trickling filter

The fundamental components of a complete trickling filter system are:

- A bed of filter medium upon which a layer of microbial slime is promoted and developed.

- An enclosure or a container which houses the bed of filter medium.

- A system for distributing the flow of wastewater over the filter medium.

- A system for removing and disposing of any sludge from the treated effluent.

The treatment of sewage or other wastewater with trickling filters is among the oldest and most well characterized treatment technologies.

A trickling filter is also often called a *trickle filter*, *trickling biofilter*, *biofilter*, *biological filter* or *biological trickling filter*.

Treatment of Other Organics

Synthetic organic materials including solvents, paints, pharmaceuticals, pesticides, products from coke production and so forth can be very difficult to treat. Treatment methods are often specific to the material being treated. Methods include advanced oxidation processing, distillation, adsorption, vitrification, incineration, chemical immobilisation or landfill disposal. Some materials such as some detergents may be capable of biological degradation and in such cases, a modified form of wastewater treatment can be used.

Treatment of Acids and Alkalis

Acids and alkalis can usually be neutralised under controlled conditions. Neutralisation frequently produces a precipitate that will require treatment as a solid residue that may also be toxic. In some cases, gases may be evolved requiring treatment for the gas stream. Some other forms of treatment are usually required following neutralisation.

Waste streams rich in hardness ions as from de-ionisation processes can readily lose the hardness ions in a buildup of precipitated calcium and magnesium salts. This precipitation process can cause severe *furring* of pipes and can, in extreme cases, cause the blockage of disposal pipes. A 1 metre diameter industrial marine discharge pipe serving a major chemicals complex was blocked by such salts in the 1970s. Treatment is by concentration of de-ionisation waste waters and disposal to landfill or by careful pH management of the released wastewater.

Treatment of Toxic Materials

Toxic materials including many organic materials, metals (such as zinc, silver, cadmium, thallium, etc.) acids, alkalis, non-metallic elements (such as arsenic or selenium) are generally resistant to biological processes unless very dilute. Metals can often be precipitated out by changing the pH or by treatment with other chemicals. Many, however, are resistant to treatment or mitigation and may require concentration followed by landfilling or recycling. Dissolved organics can be *incinerated* within the wastewater by the advanced oxidation process.

Agricultural Wastewater Treatment

Agricultural wastewater treatment is a farm management agenda for controlling pollution from surface runoff that may be contaminated by chemicals in fertiliser, pesticides, animal slurry, crop residues or irrigation water.

Riparian buffer lining a creek in Iowa

Nonpoint Source Pollution

Nonpoint source pollution from farms is caused by surface runoff from fields during rain storms. Agricultural runoff is a major source of pollution, in some cases the only source, in many watersheds.

Sediment Runoff

Soil washed off fields is the largest source of agricultural pollution in the United States. Excess sediment causes high levels of turbidity in water bodies, which can inhibit growth of aquatic plants, clog fish gills and smother animal larvae.

Highly erodible soils on a farm in Iowa

Farmers may utilize erosion controls to reduce runoff flows and retain soil on their fields. Common techniques include:

- contour ploughing
- crop mulching

- crop rotation

- planting perennial crops

- installing riparian buffers.

Nutrient Runoff

Nitrogen and phosphorus are key pollutants found in runoff, and they are applied to farmland in several ways, such as in the form of commercial fertilizer, animal manure, or municipal or industrial wastewater (effluent) or sludge. These chemicals may also enter runoff from crop residues, irrigation water, wildlife, and atmospheric deposition.

Manure spreader

Farmers can develop and implement nutrient management plans to mitigate impacts on water quality by:

- mapping and documenting fields, crop types, soil types, water bodies

- developing realistic crop yield projections

- conducting soil tests and nutrient analyses of manures and/or sludges applied

- identifying other significant nutrient sources (e.g., irrigation water)

- evaluating significant field features such as highly erodible soils, subsurface drains, and shallow aquifers

- applying fertilizers, manures, and/or sludges based on realistic yield goals and using precision agriculture techniques.

Pesticides

Pesticides are widely used by farmers to control plant pests and enhance production, but chemical pesticides can also cause water quality problems. Pesticides may appear in surface water due to:

Aerial application (crop dusting) of pesticides over a soybean field in the U.S.

- direct application (e.g. aerial spraying or broadcasting over water bodies)

- runoff during rain storms

- aerial drift (from adjacent fields).

Some pesticides have also been detected in groundwater.

Farmers may use Integrated Pest Management (IPM) techniques (which can include biological pest control) to maintain control over pests, reduce reliance on chemical pesticides, and protect water quality.

There are few safe ways of disposing of pesticide surpluses other than through containment in well managed landfills or by incineration. In some parts of the world, spraying on land is a permitted method of disposal.

Point Source Pollution

Farms with large livestock and poultry operations, such as factory farms, can be a major source of point source wastewater. In the United States, these facilities are called *concentrated animal feeding operations* or *confined animal feeding operations* and are being subject to increasing government regulation.

Animal Wastes

The constituents of animal wastewater typically contain

- Strong organic content — much stronger than human sewage

- High solids concentration

- High nitrate and phosphorus content

- Antibiotics

- Synthetic hormones

- Often high concentrations of parasites and their eggs

- Spores of *Cryptosporidium* (a protozoan) resistant to drinking water treatment processes

- Spores of *Giardia*

- Human pathogenic bacteria such as *Brucella* and *Salmonella*

Confined Animal Feeding Operation in the United States

Animal wastes from cattle can be produced as solid or semisolid manure or as a liquid slurry. The production of slurry is especially common in housed dairy cattle.

Treatment

Whilst solid manure heaps outdoors can give rise to polluting wastewaters from runoff, this type of waste is usually relatively easy to treat by containment and/or covering of the heap.

Animal slurries require special handling and are usually treated by containment in lagoons before disposal by spray or trickle application to grassland. Constructed wetlands are sometimes used to facilitate treatment of animal wastes, as are anaerobic lagoons. Excessive application or application to sodden land or insufficient land area can result in direct runoff to watercourses, with the potential for causing severe pollution. Application of slurries to land overlying aquifers can result in direct contamination or, more commonly, elevation of nitrogen levels as nitrite or nitrate.

The disposal of any wastewater containing animal waste upstream of a drinking water intake can pose serious health problems to those drinking the water because of the highly resistant spores present in many animals that are capable of causing disabling disease in humans. This risk exists even for very low-level seepage via shallow surface drains or from rainfall run-off.

Some animal slurries are treated by mixing with straws and composted at high temperature to produce a bacteriologically sterile and friable manure for soil improvement.

Piggery Waste

Piggery waste is comparable to other animal wastes and is processed as for general animal waste, except that many piggery wastes contain elevated levels of copper that can be toxic in the natural environment. The liquid fraction of the waste is frequently separated off and re-used in the piggery to avoid the prohibitively expensive costs of disposing of copper-rich liquid. Ascarid worms and their eggs are also common in piggery waste and can infect humans if wastewater treatment is ineffective.

Hog confinement barn or piggery

Silage Liquor

Fresh or wilted grass or other green crops can be made into a semi-fermented product called silage which can be stored and used as winter forage for cattle and sheep. The production of silage often involves the use of an acid conditioner such as sulfuric acid or formic acid. The process of silage making frequently produces a yellow-brown strongly smelling liquid which is very rich in simple sugars, alcohol, short-chain organic acids and silage conditioner. This liquor is one of the most polluting organic substances known. The volume of silage liquor produced is generally in proportion to the moisture content of the ensiled material.

Treatment

Silage liquor is best treated through prevention by wilting crops well before silage making. Any silage liquor that is produced can be used as part of the food for pigs. The most effective treatment is by containment in a slurry lagoon and by subsequent spreading on land following substantial dilution with slurry. Containment of silage liquor on its own can cause structural problems in concrete pits because of the acidic nature of silage liquor.

Milking Parlour (Dairy Farming) Wastes

Although milk has a deserved reputation as an important and valuable food product, its presence in wastewaters is highly polluting because of its organic strength, which can lead to very rapid de-oxygenation of receiving waters. Milking parlour wastes also contain large volumes of wash-down water, some animal waste together with cleaning and disinfection chemicals.

Treatment

Milking parlour wastes are often treated in admixture with human sewage in a local sewage treatment plant. This ensures that disinfectants and cleaning agents are sufficiently diluted and amenable to treatment. Running milking wastewaters into a farm slurry lagoon is a possible option although this tends to consume lagoon capacity very quickly. Land spreading is also a treatment option.

Slaughtering Waste

Wastewater from slaughtering activities is similar to milking parlour waste although considerably stronger in its organic composition and therefore potentially much more polluting.

Treatment

As for milking parlour waste.

Vegetable Washing Water

Washing of vegetables produces large volumes of water contaminated by soil and vegetable pieces. Low levels of pesticides used to treat the vegetables may also be present together with moderate levels of disinfectants such as chlorine.

Treatment

Most vegetable washing waters are extensively recycled with the solids removed by settlement and filtration. The recovered soil can be returned to the land.

Firewater

Although few farms plan for fires, fires are nevertheless more common on farms than on many other industrial premises. Stores of pesticides, herbicides, fuel oil for farm machinery and fertilizers can all help promote fire and can all be present in environmentally lethal quantities in firewater from fire fighting at farms.

Treatment

All farm environmental management plans should allow for containment of substantial quantities of firewater and for its subsequent recovery and disposal by specialist disposal companies. The concentration and mixture of contaminants in firewater make them unsuited to any treatment method available on the farm. Even land spreading has produced severe taste and odour problems for downstream water supply companies in the past.

References

- Tchobanoglous, George; Burton, Franklin L.; Stensel, H. David; Metcalf & Eddy, Inc. (2003). Wastewater Engineering: Treatment and Reuse (4th ed.). McGraw-Hill. ISBN 0-07-112250-8.

- Khopkar, S. M. (2004). Environmental Pollution Monitoring And Control. New Delhi: New Age International. p. 299. ISBN 81-224-1507-5.

- Martin V. Melosi (2010). The Sanitary City: Environmental Services in Urban America from Colonial Times to the Present. University of Pittsburgh Press. p. 110. ISBN 9780822973379.

- Colin A. Russell (2003). Edward Frankland: Chemistry, Controversy and Conspiracy in Victorian England. Cambridge University Press. pp. 372–380. ISBN 9780521545815.

- Sharma, Sanjay Kumar; Sanghi, Rashmi (2012). Advances in Water Treatment and Pollution Prevention. Springer. ISBN 9789400742048. Retrieved 2013-02-07.

- Tilley, David F. (2011). Aerobic Wastewater Treatment Processes: History and Development. IWA Publishing. ISBN 9781843395423. Retrieved 2013-02-07.

- Goldman, Steven J., Jackson, Katharine & Bursztynsky, Taras A. Erosion & Sediment Control Handbook. McGraw-Hill (1986). ISBN 0-07-023655-0. pp. 8.2, 8.12.

- Franson, Mary Ann. Standard Methods for the Examination of Water and Wastewater. 14th ed. (1975) APHA, AWWA & WPCF. ISBN 0-87553-078-8. pp. 89–98

- Natural Environmental Research Council – River sewage pollution found to be disrupting fish hormones. Planetearth.nerc.ac.uk. Retrieved on 2012-12-19.

Recycling: An Overview

Certain waste material can be reused and the process by which it becomes reusable is known as recycling. Recycling can be of certain types such as plastic recycling, computer recycling, ferrous metal recycling and precycling. The following chapter explains to the reader the importance of recycling.

Recycling

The three chasing arrows of the international recycling logo. It is sometimes accompanied by the text "reduce, reuse and recycle".

Recycling is the process of converting waste materials into reusable objects to prevent waste of potentially useful materials, reduce the consumption of fresh raw materials, energy usage, air pollution (from incineration) and water pollution (from landfilling) by decreasing the need for "conventional" waste disposal and lowering greenhouse gas emissions compared to plastic production. Recycling is a key component of modern waste reduction and is the third component of the "Reduce, Reuse and Recycle" waste hierarchy.

There are some ISO standards related to recycling such as ISO 15270:2008 for plastics waste and ISO 14001:2004 for environmental management control of recycling practice.

Recyclable materials include many kinds of glass, paper and cardboard, metal, plastic, tires, textiles and electronics. The composting or other reuse of biodegradable waste—such as food or garden waste—is also considered recycling. Materials to be recycled are either brought to a collection centre or picked up from the curbside, then sorted, cleaned and reprocessed into new materials destined for manufacturing.

In the strictest sense, recycling of a material would produce a fresh supply of the same material—for example, used office paper would be converted into new office paper, or used polystyrene foam into new polystyrene. However, this is often difficult or too expensive (compared with producing the same product from raw materials or other sources), so "recycling" of many products or materials involves their *reuse* in producing different materials (for example, paperboard) instead. Another form of recycling is the salvage of certain materials from complex products, either due to their intrinsic value (such as lead from car batteries, or gold from circuit boards), or due to their hazardous nature (e.g., removal and reuse of mercury from thermometers and thermostats).

History

Origins

Recycling has been a common practice for most of human history, with recorded advocates as far back as Plato in 400 BC. During periods when resources were scarce, archaeological studies of ancient waste dumps show less household waste (such as ash, broken tools and pottery)—implying more waste was being recycled in the absence of new material.

An American poster from World War II

In pre-industrial times, there is evidence of scrap bronze and other metals being collected in Europe and melted down for perpetual reuse. Paper recycling was first recorded in 1031, when Japanese shops sold repulped paper. In Britain dust and ash from wood and coal fires was collected by "dustmen" and downcycled as a base material used in brick making. The main driver for these types of recycling was the economic advantage of obtaining recycled feedstock instead of acquiring virgin material, as well as a lack of public waste removal in ever more densely populated areas. In 1813, Benjamin Law developed the process of turning rags into "shoddy" and "mungo" wool in

Batley, Yorkshire. This material combined recycled fibers with virgin wool. The West Yorkshire shoddy industry in towns such as Batley and Dewsbury, lasted from the early 19th century to at least 1914.

Industrialization spurred demand for affordable materials; aside from rags, ferrous scrap metals were coveted as they were cheaper to acquire than virgin ore. Railroads both purchased and sold scrap metal in the 19th century, and the growing steel and automobile industries purchased scrap in the early 20th century. Many secondary goods were collected, processed and sold by peddlers who scoured dumps and city streets for discarded machinery, pots, pans and other sources of metal. By World War I, thousands of such peddlers roamed the streets of American cities, taking advantage of market forces to recycle post-consumer materials back into industrial production.

Beverage bottles were recycled with a refundable deposit at some drink manufacturers in Great Britain and Ireland around 1800, notably Schweppes. An official recycling system with refundable deposits was established in Sweden for bottles in 1884 and aluminum beverage cans in 1982; the law led to a recycling rate for beverage containers of 84–99 percent depending on type, and a glass bottle can be refilled over 20 times on average.

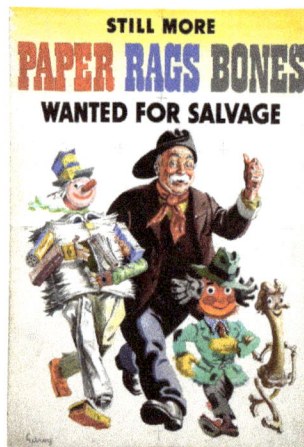

British poster from World War II

Wartime

New chemical industries created in the late 19th century both invented new materials (e.g. Bakelite (1907) and promised to transform valueless into valuable materials. Proverbially, you could not make a silk purse of a sow's ear—until the US firm Arhur D. Little published in 1921 "On the Making of Silk Purses from Sows' Ears", its research proving that when "chemistry puts on overalls and gets down to business . . .new values appear. New and better paths are opened to reach the goals desired."

Recycling was a highlight throughout World War II. During the war, financial constraints and significant material shortages due to war efforts made it necessary for countries to reuse goods and recycle materials. These resource shortages caused by the world wars, and other such world-changing occurrences, greatly encouraged recycling. The struggles of war claimed much of the material resources available, leaving little for the civilian population. It became necessary for most homes to recycle their waste, as recycling offered an extra source of materials allowing people to make the most of what was available to them. Recycling household materials meant more resources for war efforts and a better chance of victory. Massive government promotion campaigns were carried out in the home front during World War II in every country involved in the war, urging citizens to donate metals and conserve fiber, as a matter of patriotism.

Post-war

A considerable investment in recycling occurred in the 1970s, due to rising energy costs. Recycling aluminum uses only 5% of the energy required by virgin production; glass, paper and metals have less dramatic but very significant energy savings when recycled feedstock is used.

Although consumer electronics such as the television have been popular since the 1920s, recycling of them was almost unheard of until early 1991. The first electronic waste recycling scheme was implemented in Switzerland, beginning with collection of old refrigerators but gradually expanding to cover all devices. After these schemes were set up, many countries did not have the capacity to deal with the sheer quantity of e-waste they generated or its hazardous nature. They began to export the problem to developing countries without enforced environmental legislation. This is cheaper, as recycling computer monitors in the United States costs 10 times more than in China. Demand in Asia for electronic waste began to grow when scrap yards found that they could extract valuable substances such as copper, silver, iron, silicon, nickel and gold, during the recycling process. The 2000s saw a large increase in both the sale of electronic devices and their growth as a waste stream: in 2002, e-waste grew faster than any other type of waste in the EU. This caused investment in modern, automated facilities to cope with the influx of redundant appliances, especially after strict laws were implemented in 2003.

As of 2014, the European Union has about 50% of world share of the waste and recycling industries, with over 60,000 companies employing 500,000 persons, with a turnover of €24 billion. Countries have to reach recycling rates of at least 50%, while the lead countries are around 65% and the EU average is 39% as of 2013.

Legislation

Supply

For a recycling program to work, having a large, stable supply of recyclable material is crucial. Three legislative options have been used to create such a supply: mandatory recycling collection, container deposit legislation and refuse bans. Mandatory collection laws set recycling targets for cities to aim for, usually in the form that a certain percentage of a material must be diverted from the city's waste stream by a target date. The city is then responsible for working to meet this target.

Container deposit legislation involves offering a refund for the return of certain containers, typically glass, plastic and metal. When a product in such a container is purchased, a small surcharge is added to the price. This surcharge can be reclaimed by the consumer if the container is returned to a collection point. These programs have been very successful, often resulting in an 80 percent recycling rate. Despite such good results, the shift in collection costs from local government to industry and consumers has created strong opposition to the creation of such programs in some areas. A variation on this is where the manufacturer bears responsibility for the recycling of their goods. In the European Union, the WEEE Directive requires producers of consumer electronics to reimburse the recyclers' costs.

An alternative way to increase supply of recyclates is to ban the disposal of certain materials as waste, often including used oil, old batteries, tires and garden waste. One aim of this method is to create a viable economy for proper disposal of banned products. Care must be taken that enough of these recycling services exist, or such bans simply lead to increased illegal dumping.

Government-mandated Demand

Legislation has also been used to increase and maintain a demand for recycled materials. Four methods of such legislation exist: minimum recycled content mandates, utilization rates, procurement policies and recycled product labeling.

Both minimum recycled content mandates and utilization rates increase demand directly by forcing manufacturers to include recycling in their operations. Content mandates specify that a certain percentage of a new product must consist of recycled material. Utilization rates are a more flexible option: industries are permitted to meet the recycling targets at any point of their operation or even contract recycling out in exchange for tradeable credits. Opponents to both of these methods point to the large increase in reporting requirements they impose, and claim that they rob industry of necessary flexibility.

Governments have used their own purchasing power to increase recycling demand through what are called "procurement policies." These policies are either "set-asides," which reserve a certain amount of spending solely towards recycled products, or "price preference" programs which provide a larger budget when recycled items are purchased. Additional regulations can target specific cases: in the United States, for example, the Environmental Protection Agency mandates the purchase of oil, paper, tires and building insulation from recycled or re-refined sources whenever possible.

The final government regulation towards increased demand is recycled product labeling. When producers are required to label their packaging with amount of recycled material in the product (including the packaging), consumers are better able to make educated choices. Consumers with sufficient buying power can then choose more environmentally conscious options, prompt producers to increase the amount of recycled material in their products, and indirectly increase demand. Standardized recycling labeling can also have a positive effect on supply of recyclates if the labeling includes information on how and where the product can be recycled.

Recyclates

Glass recovered by crushing only one kind of beer bottle

Recyclate is a raw material that is sent to, and processed in a waste recycling plant or materials recovery facility which will be used to form new products. The material is collected in various methods and delivered to a facility where it undergoes re-manufacturing so that it can used in the production of new materials or products. For example, plastic bottles that are collected can be re-used and made into plastic pellets, a new product.

Quality of Recyclate

The quality of recyclates is recognized as one of the principal challenges that needs to be addressed for the success of a long-term vision of a green economy and achieving zero waste. Recyclate quality is generally referring to how much of the raw material is made up of target material compared to the amount of non-target material and other non-recyclable material. Only target material is likely to be recycled, so a higher amount of non-target and non-recyclable material will reduce the quantity of recycling product. A high proportion of non-target and non-recyclable material can make it more difficult for re-processors to achieve "high-quality" recycling. If the recyclate is of poor quality, it is more likely to end up being down-cycled or, in more extreme cases, sent to other recovery options or landfilled. For example, to facilitate the re-manufacturing of clear glass products there are tight restrictions for colored glass going into the re-melt process.

The quality of recyclate not only supports high quality recycling, but it can also deliver significant environmental benefits by reducing, reusing and keeping products out of landfills. High quality recycling can help support growth in the economy by maximizing the economic value of the waste material collected. Higher income levels from the sale of quality recyclates can return value which can be significant to local governments, households and businesses. Pursuing high quality recycling can also provide consumer and business confidence in the waste and resource management sector and may encourage investment in that sector.

There are many actions along the recycling supply chain that can influence and affect the material quality of recyclate. It begins with the waste producers who place non-target and non-recyclable wastes in recycling collection. This can affect the quality of final recyclate streams or require further efforts to discard those materials at later stages in the recycling process. The different collection systems can result in different levels of contamination. Depending on which materials are collected together, extra effort is required to sort this material back into separate streams and can significantly reduce the quality of the final product. Transportation and the compaction of materials can make it more difficult to separate material back into separate waste streams. Sorting facilities are not one hundred per cent effective in separating materials, despite improvements in technology and quality recyclate which can see a loss in recyclate quality. The storage of materials outside where the product can become wet can cause problems for re-processors. Reprocessing facilities may require further sorting steps to further reduce the amount of non-target and non-recyclable material. Each action along the recycling path plays a part in the quality of recyclate.

Quality Recyclate Action Plan (Scotland)

The Recyclate Quality Action Plan of Scotland sets out a number of proposed actions that the Scottish Government would like to take forward in order to drive up the quality of the materials being collected for recycling and sorted at materials recovery facilities before being exported or sold on to the reprocessing market.

The plan's objectives are to:

- Drive up the quality of recyclate.

- Deliver greater transparency about the quality of recyclate.

- Provide help to those contracting with materials recycling facilities to identify what is required of them

- Ensure compliance with the Waste (Scotland) regulations 2012.

- Stimulate a household market for quality recyclate.

- Address and reduce issues surrounding the Waste Shipment Regulations.

The plan focuses on three key areas, with fourteen actions which were identified to increase the quality of materials collected, sorted and presented to the processing market in Scotland.

The three areas of focus are:

1. Collection systems and input contamination

2. Sorting facilities – material sampling and transparency

3. Material quality benchmarking and standards

Recycling Consumer Waste

Collection

A number of different systems have been implemented to collect recyclates from the general waste stream. These systems lie along the spectrum of trade-off between public convenience and government ease and expense. The three main categories of collection are "drop-off centers," "buy-back centers" and "curbside collection."

A three-sided bin at a railway station in Germany, intended to separate paper *(left)* and plastic wrappings *(right)* from other waste *(back)*

Curbside Collection

Curbside collection encompasses many subtly different systems, which differ mostly on where in the process the recyclates are sorted and cleaned. The main categories are mixed waste collection, commingled recyclables and source separation. A waste collection vehicle generally picks up the waste.

At one end of the spectrum is mixed waste collection, in which all recyclates are collected mixed in with the rest of the waste, and the desired material is then sorted out and cleaned at a central sorting facility. This results in a large amount of recyclable waste, paper especially, being too soiled to reprocess, but has advantages as well: the city need not pay for a separate collection of recyclates and no public education is needed. Any changes to which materials are recyclable is easy to accommodate as all sorting happens in a central location.

A recycling truck collecting the contents of a recycling bin in Canberra, Australia

In a commingled or single-stream system, all recyclables for collection are mixed but kept separate from other waste. This greatly reduces the need for post-collection cleaning but does require public education on what materials are recyclable.

Source separation is the other extreme, where each material is cleaned and sorted prior to collection. This method requires the least post-collection sorting and produces the purest recyclates, but incurs additional operating costs for collection of each separate material. An extensive public education program is also required, which must be successful if recyclate contamination is to be avoided.

Source separation used to be the preferred method due to the high sorting costs incurred by commingled (mixed waste) collection. Advances in sorting technology however, have lowered this overhead substantially—many areas which had developed source separation programs have since switched to co-mingled collec-tion.

Buy-back Centers

Buy-back centers differ in that the cleaned recyclates are purchased, thus providing a clear incentive for use and creating a stable supply. The post-processed material can then be sold on, hopefully creating a profit. Unfortunately, government subsidies are necessary to make buy-back centres a viable enterprise, as according to the U.S. National Waste & Recycling Association, it costs on average US$50 to process a ton of material, which can only be resold for US$30.

Drop-off Centers

Drop-off centers require the waste producer to carry the recyclates to a central location, either an installed or mobile collection station or the reprocessing plant itself. They are the easiest type of collection to establish, but suffer from low and unpredictable throughput.

Distributed Recycling

For some waste materials such as plastic, recent technical devices called recyclebots enable a form of distributed recycling. Preliminary life-cycle analysis (LCA) indicates that such distributed recycling of HDPE to make filament of 3-D printers in rural regions is energetically favorable to either using virgin resin or conventional recycling processes because of reductions in transportation energy.

Sorting

Once commingled recyclates are collected and delivered to a central collection facility, the different types of materials must be sorted. This is done in a series of stages, many of which involve automated processes such that a truckload of material can be fully sorted in less than an hour. Some plants can now sort the materials automatically, known as single-stream recycling. In plants, a variety of materials are sorted such as paper, different types of plastics, glass, metals, food scraps and most types of batteries. A 30 percent increase in recycling rates has been seen in the areas where these plants exist.

Recycling sorting facility and processes

Initially, the commingled recyclates are removed from the collection vehicle and placed on a conveyor belt spread out in a single layer. Large pieces of corrugated fiberboard and plastic bags are removed by hand at this stage, as they can cause later machinery to jam.

Early sorting of recyclable materials: glass and plastic bottles in Poland

Next, automated machinery such as disk screens and air classifiers separate the recyclates by weight, splitting lighter paper and plastic from heavier glass and metal. Cardboard is removed from the mixed paper and the most common types of plastic, PET (#1) and HDPE (#2), are collected. This separation is usually done by hand but has become automated in some sorting centers: a spectroscopic scanner is used to differentiate between different types of paper and plastic based on the absorbed wavelengths, and subsequently divert each material into the proper collection channel.

Strong magnets are used to separate out ferrous metals, such as iron, steel and tin cans. Non-ferrous metals are ejected by magnetic eddy currents in which a rotating magnetic field induces an electric current around the aluminum cans, which in turn creates a magnetic eddy current inside the cans. This magnetic eddy current is repulsed by a large magnetic field, and the cans are ejected from the rest of the recyclate stream.

Finally, glass is sorted according to its color: brown, amber, green or clear. It may either be sorted by hand, or via an automated machine that uses colored filters to detect different colors. Glass fragments smaller than 10 millimetres (0.39 in) across cannot be sorted automatically, and are mixed together as "glass fines."

A recycling point in New Byth, Scotland, with separate containers for paper, plastics and differently colored glass

This process of recycling as well as reusing the recycled material has proven advantageous because it reduces amount of waste sent to landfills, conserves natural resources, saves energy, reduces greenhouse gas emissions and helps create new jobs. Recycled materials can also be converted into new products that can be consumed again, such as paper, plastic and glass.

The City and County of San Francisco's Department of the Environment is attempting to achieve a citywide goal of Zero Waste by 2020. San Francisco's refuse hauler, Recology, operates an effective recyclables sorting facility in San Francisco, which helped San Francisco reach a record-breaking diversion rate of 80%.

Rinsing

Food packaging should no longer contain any organic matter (organic matter, if any, needs to be placed in a biodegradable waste bin or be buried in a garden). Since no trace of biodegradable material is best kept in the packaging before placing it in a trash bag, some packaging also needs to be rinsed.

Recycling Industrial Waste

Although many government programs are concentrated on recycling at home, a 64% of waste in the United Kingdom is generated by industry. The focus of many recycling programs done by industry is the cost–effectiveness of recycling. The ubiquitous nature of cardboard packaging makes cardboard a commonly recycled waste product by companies that deal heavily in packaged goods, like retail stores, warehouses and distributors of goods. Other industries deal in niche or specialized products, depending on the nature of the waste materials that are present.

The glass, lumber, wood pulp and paper manufacturers all deal directly in commonly recycled materials; however, old rubber tires may be collected and recycled by independent tire dealers for a profit.

Mounds of shredded rubber tires are ready for processing

Levels of metals recycling are generally low. In 2010, the International Resource Panel, hosted by the United Nations Environment Programme (UNEP) published reports on metal stocks that exist within society and their recycling rates. The Panel reported that the increase in the use of metals during the 20th and into the 21st century has led to a substantial shift in metal stocks from below ground to use in applications within society above ground. For example, the in-use stock of copper in the USA grew from 73 to 238 kg per capita between 1932 and 1999.

The report authors observed that, as metals are inherently recyclable, the metal stocks in society can serve as huge mines above ground (the term "**urban mining**" has been coined with this idea in mind). However, they found that the recycling rates of many metals are very low. The report warned that the recycling rates of some rare metals used in applications such as mobile phones, battery packs for hybrid cars and fuel cells, are so low that unless future end-of-life recycling rates are dramatically stepped up these critical metals will become unavailable for use in modern technology.

Aerial photo of a ship recycling facility in Chittagong, Bangladesh

The military recycles some metals. The U.S. Navy's Ship Disposal Program uses ship breaking to reclaim the steel of old vessels. Ships may also be sunk to create an artificial reef. Uranium is a very dense metal that has qualities superior to lead and titanium for many military and industrial uses. The uranium left over from processing it into nuclear weapons and fuel for nuclear reactors is called depleted uranium, and it is used by all branches of the U.S. military use for armour-piercing shells and shielding.

The construction industry may recycle concrete and old road surface pavement, selling their waste materials for profit.

Some industries, like the renewable energy industry and solar photovoltaic technology in particular, are being proactive in setting up recycling policies even before there is considerable volume to their waste streams, anticipating future demand during their rapid growth.

Recycling of plastics is more difficult, as most programs are not able to reach the necessary level of quality. Recycling of PVC often results in downcycling of the material, which means only products of lower quality standard can be made with the recycled material. A new approach which allows an equal level of quality is the Vinyloop process. It was used after the London Olympics 2012 to fulfill the PVC Policy.

E-waste Recycling

E-waste is a growing problem, accounting for 20-50 million metric tons of global waste per year according to the EPA. It is also the fastest growing waste stream in the EU. Many recyclers do not recycle e-waste responsibly. After the cargo barge Khian Sea dumped 14,000 metric tons of toxic ash in Haiti, the Basel Convention was formed to stem the flow of hazardous substances into poorer countries. They created the e-Stewards certification to ensure that recyclers are held to the highest standards for environmental responsibility and to help consumers identify responsible recyclers. This works alongside other prominent legislation, such as the Waste Electrical and Electronic Equipment Directive of the EU the United States National Computer Recycling Act, to prevent poisonous chemicals from entering waterways and the atmosphere.

Microprocessors retrieved from waste stream

In the recycling process, television sets, monitors, cell phones and computers are typically tested for reuse and repaired. If broken, they may be disassembled for parts still having high value if labor is cheap enough. Other e-waste is shredded to pieces roughly 10 centimetres (3.9 in) in size, and manually checked to separate out toxic batteries and capacitors which contain poisonous metals. The remaining pieces are further shredded to 10 millimetres (0.39 in) particles and passed under a magnet to remove ferrous metals. An eddy current ejects non-ferrous metals, which are sorted by density either by a centrifuge or vibrating plates. Precious metals can be dissolved in acid, sorted, and smelted into ingots. The remaining glass and plastic fractions are separated by density and sold to re-processors. Television sets and monitors must be manually disassembled to remove lead from CRTs or the mercury backlight from LCDs.

Plastic Recycling

Plastic recycling is the process of recovering scrap or waste plastic and reprocessing the material into useful products, sometimes completely different in form from their original state. For instance, this could mean melting down soft drink bottles and then casting them as plastic chairs and tables.

A container for recycling used plastic spoons into material for 3D printing

Physical Recycling

Some plastics are remelted to form new plastic objects; for example, PET water bottles can be converted into polyester destined for clothing. A disadvantage of this type of recycling is that the molecular weight of the polymer can change further and the levels of unwanted substances in the plastic can increase with each remelt.

Chemical Recycling

For some polymers, it is possible to convert them back into monomers, for example PET can be treated with an alcohol and a catalyst to form a dialkyl terephthalate. The terephthalate diester can be used with ethylene glycol to form a new polyester polymer, thus making it possible to use the pure polymer again.

Waste Plastic Pyrolysis to Fuel Oil

Another process involves conversion of assorted polymers into petroleum by a much less precise thermal depolymerization process. Such a process would be able to accept almost any polymer or mix of polymers, including thermoset materials such as vulcanized rubber tires and the biopolymers in feathers and other agricultural waste. Like natural petroleum, the chemicals produced can be used as fuels or as feedstock. A RESEM Technology plant of this type in Carthage, Missouri, USA, uses turkey waste as input material. Gasification is a similar process, but is not technically recycling since polymers are not likely to become the result. Plastic Pyrolysis can convert petroleum based waste streams such as plastics into quality fuels, carbons. Given below is the list of suitable plastic raw materials for pyrolysis:

- Mixed plastic (HDPE, LDPE, PE, PP, Nylon, Teflon, PS, ABS, FRP, etc.)

- Mixed waste plastic from waste paper mill

- Multi-layered plastic

Recycling Codes

In order to meet recyclers' needs while providing manufacturers a consistent, uniform system, a coding system was developed. The recycling code for plastics was introduced in 1988 by the plastics industry through the Society of the Plastics Industry. Because municipal recycling programs traditionally have targeted packaging—primarily bottles and containers—the resin coding system offered a means of identifying the resin content of bottles and containers commonly found in the residential waste stream.

Recycling codes on products

Plastic products are printed with numbers 1–7 depending on the type of resin. Type 1 (polyethylene terephthalate) is commonly found in soft drink and water bottles. Type 2 (high-density polyethylene) is found in most hard plastics such as milk jugs, laundry detergent bottles and some dishware. Type 3 (polyvinyl chloride) includes items such as shampoo bottles, shower curtains, hula hoops, credit cards, wire jacketing, medical equipment, siding and piping. Type 4 (low-density polyethylene) is found in shopping bags, squeezable bottles, tote bags, clothing, furniture and carpet. Type 5 is polypropylene and makes up syrup bottles, straws, Tupperware and some automotive parts. Type 6 is polystyrene and makes up meat trays, egg cartons, clamshell containers and compact disc cases. Type 7 includes all other plastics such as bulletproof materials, 3- and 5-gallon water bottles and sunglasses. Having a recycling code or the chasing arrows logo on a material is not an automatic indicator that a material is recyclable but rather an explanation of what the material is. Types 1 and 2 are the most commonly recycled.

Economic Impact

Critics dispute the net economic and environmental benefits of recycling over its costs, and suggest that proponents of recycling often make matters worse and suffer from confirmation bias. Specifically, critics argue that the costs and energy used in collection and transportation detract from (and outweigh) the costs and energy saved in the production process; also that the jobs produced by the recycling industry can be a poor trade for the jobs lost in logging, mining, and other industries associated with production; and that materials such as paper pulp can only be recycled a few times before material degradation prevents further recycling.

The National Waste and Recycling Association (NWRA), reported in May 2015, that recycling and waste made a $6.7 billion economic impact in Ohio, U.S., and employed 14,000 people.

Cost–benefit Analysis

Environmental effects of recycling		
Material	**Energy savings**	**Air pollution savings**
Aluminium	95%	95%
Cardboard	24%	–
Glass	5–30%	20%
Paper	40%	73%
Plastics	70%	–
Steel	60%	–

There is some debate over whether recycling is economically efficient. It is said that dumping 10,000 tons of waste in a landfill creates six jobs, while recycling 10,000 tons of waste can create over 36 jobs. However, the cost effectiveness of creating the additional jobs remains unproven. According to the U.S. Recycling Economic Informational Study, there are over 50,000 recycling establishments that have created over a million jobs in the US. Two years after New York City declared that implementing recycling programs would be "a drain on the city," New York City leaders realized that an efficient recycling system could save the city over $20 million. Municipalities often see fiscal benefits from implementing recycling programs, largely due to the reduced landfill costs. A study conducted by the Technical University of Denmark according to the Economist found that in 83 percent of cases, recycling is the most efficient method to dispose of household waste. However, a 2004 assessment by the Danish Environmental Assessment Institute concluded that incineration was the most effective method for disposing of drink containers, even aluminium ones.

Fiscal efficiency is separate from economic efficiency. Economic analysis of recycling does not include what economists call externalities, which are unpriced costs and benefits that accrue to individuals outside of private transactions. Examples include: decreased air pollution and greenhouse gases from incineration, reduced hazardous waste leaching from landfills, reduced energy consumption, and reduced waste and resource consumption, which leads to a reduction in environmentally damaging mining and timber activity. About 4,000 minerals are known, of these only a few hundred minerals in the world are relatively common. Known reserves of phosphorus will be exhausted within the next 100 years at current rates of usage. Without mechanisms such as taxes or subsidies to internalize externalities, businesses will ignore them despite the costs imposed on society. To make such nonfiscal benefits economically relevant, advocates have pushed for legislative action to increase the demand for recycled materials. The United States Environmental Protection Agency (EPA) has concluded in favor of recycling, saying that recycling efforts reduced the country's carbon emissions by a net 49 million metric tonnes in 2005. In the United Kingdom, the Waste and Resources Action Programme stated that Great Britain's recycling efforts reduce CO_2 emissions by 10–15 million tonnes a year. Recycling is more efficient in densely populated areas, as there are economies of scale involved.

Certain requirements must be met for recycling to be economically feasible and environmentally effective. These include an adequate source of recyclates, a system to extract those recyclates from the waste stream, a nearby factory capable of reprocessing the recyclates, and a potential demand for the recycled products. These last two requirements are often overlooked—without both an industrial market for production using the collected materials and a consumer market for the manufactured goods, recycling is incomplete and in fact only "collection".

Free-market economist Julian Simon remarked "There are three ways society can organize waste disposal: (a) commanding,(b) guiding by tax and subsidy, and (c) leaving it to the individual and the market". These principles appear to divide economic thinkers today.

Frank Ackerman favours a high level of government intervention to provide recycling services. He believes that recycling's benefit cannot be effectively quantified by traditional *laissez-faire* economics. Allen Hershkowitz supports intervention, saying that it is a public service equal to education and policing. He argues that manufacturers should shoulder more of the burden of waste disposal.

Wrecked automobiles gathered for smelting

Paul Calcott and Margaret Walls advocate the second option. A deposit refund scheme and a small refuse charge would encourage recycling but not at the expense of fly-tipping. Thomas C. Kinnaman concludes that a landfill tax would force consumers, companies and councils to recycle more.

Most free-market thinkers detest subsidy and intervention because they waste resources. Terry Anderson and Donald Leal think that all recycling programmes should be privately operated, and therefore would only operate if the money saved by recycling exceeds its costs. Daniel K. Benjamin argues that it wastes people's resources and lowers the wealth of a population.

Trade in Recyclates

Certain countries trade in unprocessed recyclates. Some have complained that the ultimate fate of recyclates sold to another country is unknown and they may end up in landfills instead of reprocessed. According to one report, in America, 50–80 percent of computers destined for recycling are actually not recycled. There are reports of illegal-waste imports to China being dismantled and recycled solely for monetary gain, without consideration for workers' health or environmental damage. Although the Chinese government has banned these practices, it has not been able to eradicate them. In 2008, the prices of recyclable waste plummeted before rebounding in 2009. Cardboard averaged about £53/tonne from 2004–2008, dropped to £19/tonne, and then went up to £59/tonne in May 2009. PET plastic averaged about £156/tonne, dropped to £75/tonne and then moved up to £195/tonne in May 2009.

Certain regions have difficulty using or exporting as much of a material as they recycle. This problem is most prevalent with glass: both Britain and the U.S. import large quantities of wine bottled in green glass. Though much of this glass is sent to be recycled, outside the American Midwest there is not enough wine production to use all of the reprocessed material. The extra must be downcycled into building materials or re-inserted into the regular waste stream.

Similarly, the northwestern United States has difficulty finding markets for recycled newspaper, given the large number of pulp mills in the region as well as the proximity to Asian markets. In other areas of the U.S., however, demand for used newsprint has seen wide fluctuation.

In some U.S. states, a program called RecycleBank pays people to recycle, receiving money from local municipalities for the reduction in landfill space which must be purchased. It uses a single stream process in which all material is automatically sorted.

Criticisms and Responses

Much of the difficulty inherent in recycling comes from the fact that most products are not designed with recycling in mind. The concept of sustainable design aims to solve this problem, and was laid out in the book *Cradle to Cradle: Remaking the Way We Make Things* by architect William McDonough and chemist Michael Braungart. They suggest that every product (and all packaging they require) should have a complete "closed-loop" cycle mapped out for each component—a way in which every component will either return to the natural ecosystem through biodegradation or be recycled indefinitely.

Complete recycling is impossible from a practical standpoint. In summary, substitution and recycling strategies only delay the depletion of non-renewable stocks and therefore may buy time in the transition to true or strong sustainability, which ultimately is only guaranteed in an economy based on renewable resources.

—M. H. Huesemann, 2003

While recycling diverts waste from entering directly into landfill sites, current recycling misses the dissipative components. Complete recycling is impracticable as highly dispersed wastes become so diluted that the energy needed for their recovery becomes increasingly excessive. "For example, how will it ever be possible to recycle the numerous chlorinated organic hydrocarbons that have bioaccumulated in animal and human tissues across the globe, the copper dispersed in fungicides, the lead in widely applied paints, or the zinc oxides present in the finely dispersed rubber powder that is abraded from automobile tires?"

As with environmental economics, care must be taken to ensure a complete view of the costs and benefits involved. For example, paperboard packaging for food products is more easily recycled than most plastic, but is heavier to ship and may result in more waste from spoilage.

Energy and Material Flows

The amount of energy saved through recycling depends upon the material being recycled and the type of energy accounting that is used. Correct accounting for this saved energy can be accomplished with life-cycle analysis using real energy values. In addition, exergy, which is a measure of useful energy can be used. In general, it takes far less energy to produce a unit mass of recycled materials than it does to make the same mass of virgin materials.

Some scholars use emergy (spelled with an m) analysis, for example, budgets for the amount of energy of one kind (exergy) that is required to make or transform things into another kind of product or service. Emergy calculations take into account economics which can alter pure physics based results. Using emergy life-cycle analysis researchers have concluded that materials with large refining costs have the greatest potential for high recycle benefits. Moreover, the highest emergy efficiency accrues from systems geared toward material recycling, where materials are engineered to recycle back into their original form and purpose, followed by adaptive reuse systems where the materials are recycled into a different kind of product, and then by-product reuse systems where parts of the products are used to make an entirely different product.

The Energy Information Administration (EIA) states on its website that "a paper mill uses 40 percent less energy to make paper from recycled paper than it does to make paper from fresh lumber." Some critics argue that it takes more energy to produce recycled products than it does to dispose of them in traditional landfill methods, since the curbside collection of recyclables often requires a second waste truck. However, recycling proponents point out that a second timber or logging truck is eliminated when paper is collected for recycling, so the net energy consumption is the same. An Emergy life-cycle analysis on recycling revealed that fly ash, aluminum, recycled concrete aggregate, recycled plastic, and steel yield higher efficiency ratios, whereas the recycling of lumber generates the lowest recycle benefit ratio. Hence, the specific nature of the recycling process, the methods used to analyse the process, and the products involved affect the energy savings budgets.

It is difficult to determine the amount of energy consumed or produced in waste disposal processes in broader ecological terms, where causal relations dissipate into complex networks of material and energy flow. For example, "cities do not follow all the strategies of ecosystem development. Biogeochemical paths become fairly straight relative to wild ecosystems, with very reduced recycling, resulting in large flows of waste and low total energy efficiencies. By contrast, in wild ecosystems, one population's wastes are another population's resources, and succession results in efficient exploitation of available resources. However, even modernized cities may still be in the earliest stages of a succession that may take centuries or millennia to complete." How much energy is used in recycling also depends on the type of material being recycled and the process used to do so. Aluminium is generally agreed to use far less energy when recycled rather than being produced from scratch. The EPA states that "recycling aluminum cans, for example, saves 95 percent of the energy required to make the same amount of aluminum from its virgin source, bauxite." In 2009 more than half of all aluminium cans produced came from recycled aluminium.

Every year, millions of tons of materials are being exploited from the earth's crust, and processed into consumer and capital goods. After decades to centuries, most of these materials are "lost". With the exception of some pieces of art or religious relics, they are no longer engaged in the consumption process. Where are they? Recycling is only an intermediate solution for such materials, although it does prolong the residence time in the anthroposphere. For thermodynamic reasons, however, recycling cannot prevent the final need for an ultimate sink.

— P. H. Brunner

Economist Steven Landsburg has suggested that the sole benefit of reducing landfill space is trumped by the energy needed and resulting pollution from the recycling process. Others, however, have calculated through life-cycle assessment that producing recycled paper uses less energy and water than harvesting, pulping, processing, and transporting virgin trees. When less recycled paper is used, additional energy is needed to create and maintain farmed forests until these forests are as self-sustainable as virgin forests.

Other studies have shown that recycling in itself is inefficient to perform the "decoupling" of economic development from the depletion of non-renewable raw materials that is necessary for sustainable development. The international transportation or recycle material flows through "... different trade networks of the three countries result in different flows, decay rates, and potential recycling returns." As global consumption of a natural resources grows, its depletion is inevitable. The best recycling can do is to delay, complete closure of material loops to achieve 100 percent recycling of nonrenewables is impossible as micro-trace materials dissipate into the environment causing severe damage to the planet's ecosystems. Historically, this was identified as the metabolic rift by Karl Marx, who identified the unequal exchange rate between energy and nutrients flowing from rural areas to feed urban cities that create effluent wastes degrading the planet's ecological capital, such as loss in soil nutrient production. Energy conservation also leads to what is known as Jevon's paradox, where improvements in energy efficiency lowers the cost of production and leads to a rebound effect where rates of consumption and economic growth increases.

A shop in New York only sells items recycled from demolished buildings

Costs

The amount of money actually saved through recycling depends on the efficiency of the recycling program used to do it. The Institute for Local Self-Reliance argues that the cost of recycling depends on various factors, such as landfill fees and the amount of disposal that the community recycles. It states that communities begin to save money when they treat recycling as a replacement for their traditional waste system rather than an add-on to it and by "redesigning their collection schedules and/or trucks."

In some cases, the cost of recyclable materials also exceeds the cost of raw materials. Virgin plastic resin costs 40 percent less than recycled resin. Additionally, a United States Environmental Protection Agency (EPA) study that tracked the price of clear glass from July 15 to August 2, 1991, found that the average cost per ton ranged from $40 to $60, while a USGS report shows that the cost per ton of raw silica sand from years 1993 to 1997 fell between $17.33 and $18.10.

In 1996 and 2015 articles for *The New York Times*, John Tierney argued that it costs more money to recycle the trash of New York City than it does to dispose of it in a landfill. Tierney argued that the recycling process employs people to do the additional waste disposal, sorting, inspecting, and many fees are often charged because the processing costs used to make the end product are often more than the profit from its sale. Tierney also referenced a study conducted by the Solid Waste Association of North America (SWANA) that found in the six communities involved in the study, "all but one of the curbside recycling programs, and all the composting operations and waste-to-energy incinerators, increased the cost of waste disposal."

Tierney also points out that "the prices paid for scrap materials are a measure of their environmental value as recyclables. Scrap aluminum fetches a high price because recycling it consumes so much less energy than manufacturing new aluminum."

However, comparing the market cost of recyclable material with the cost of new raw materials ignores economic externalities—the costs that are currently not counted by the market. Creating a new piece of plastic, for instance, may cause more pollution and be less sustainable than recycling a similar piece of plastic, but these factors will not be counted in market cost. A life cycle assessment can be used to determine the levels of externalities and decide whether the recycling may be worthwhile despite unfavorable market costs. Alternatively, legal means (such as a carbon tax) can be used to bring externalities into the market, so that the market cost of the material becomes close to the true cost.

In a 2007 article, Michael Munger, chairman of political science at Duke University, wrote that "if recycling is more expensive than using new materials, it can't possibly be efficient.... There is a simple test for determining whether something is a resource... or just garbage... If someone will pay you for the item, it's a resource.... But if you have to pay someone to take the item away,... then the item is garbage."

In a 2002 article for The Heartland Institute, Jerry Taylor, director of natural resource studies at the Cato Institute, wrote, "If it costs X to deliver newly manufactured plastic to the market, for example, but it costs 10X to deliver reused plastic to the market, we can conclude the resources required to recycle plastic are 10 times more scarce than the resources required to make plastic from scratch. And because recycling is supposed to be about the conservation of resources, mandating recycling under those circumstances will do more harm than good."

Working Conditions

The recycling of waste electrical and electronic equipment in India and China generates a significant amount of pollution. Informal recycling in an underground economy of these countries has generated an environmental and health disaster. High levels of lead (Pb), polybrominated diphenylethers (PBDEs), polychlorinated dioxins and furans, as well as polybrominated dioxins and furans (PCDD/Fs and PBDD/Fs) concentrated in the air, bottom ash, dust, soil, water and sediments in areas surrounding recycling sites. Critics also argue that while recycling may create jobs, they are often jobs with low wages and terrible working conditions. These jobs are sometimes considered to be make-work jobs that don't produce as much as the cost of wages to pay for those jobs. In areas without many environmental regulations and/or worker protections, jobs involved in recycling such as ship breaking can result in deplorable conditions for both workers and the surrounding communities.

People in Brazil who earn their living by collecting and sorting garbage and selling them for recycling

Environmental Impact

Economist Steven Landsburg, author of a paper entitled "Why I Am Not an Environmentalist," claimed that paper recycling actually reduces tree populations. He argues that because paper companies have incentives to replenish their forests, large demands for paper lead to large forests, while reduced demand for paper leads to fewer "farmed" forests.

When foresting companies cut down trees, more are planted in their place. Most paper comes from pulp forests grown specifically for paper production. Many environmentalists point out, however, that "farmed" forests are inferior to virgin forests in several ways.

Farmed forests are not able to fix the soil as quickly as virgin forests, causing widespread soil erosion and often requiring large amounts of fertilizer to maintain while containing little tree and wild-life biodiversity compared to virgin forests. Also, the new trees planted are not as big as the trees that were cut down, and the argument that there will be "more trees" is not compelling to forestry advocates when they are counting saplings.

In particular, wood from tropical rainforests is rarely harvested for paper because of their heterogeneity. According to the United Nations Framework Convention on Climate Change secretariat, the overwhelming direct cause of deforestation is subsistence farming (48% of deforestation) and commercial agriculture (32%), which is linked to food, not paper production.

Possible Income Loss and Social Costs

In some countries, recycling is performed by the entrepreneurial poor such as the karung guni, zabbaleen, the rag-and-bone man, waste picker, and junk man. With the creation of large recycling organizations that may be profitable, either by law or economies of scale, the poor are more likely to be driven out of the recycling and the re-manufacturing market. To compensate for this loss of income, a society may need to create additional forms of societal programs to help support the poor. Like the parable of the broken window, there is a net loss to the poor and possibly the whole of a society to make recycling artificially profitable e.g. through the law. However, in Brazil and Argentina, waste pickers/informal recyclers work alongside the authorities, in fully or semi-funded cooperatives, allowing informal recycling to be legitimized as a paid public sector job.

Because the social support of a country is likely to be less than the loss of income to the poor undertaking recycling, there is a greater chance the poor will come in conflict with the large recycling organizations. This means fewer people can decide if certain waste is more economically reusable in its current form rather than being reprocessed. Contrasted to the recycling poor, the efficiency of their recycling may actually be higher for some materials because individuals have greater control over what is considered "waste."

One labor-intensive underused waste is electronic and computer waste. Because this waste may still be functional and wanted mostly by those on lower incomes, who may sell or use it at a greater efficiency than large recyclers.

Some recycling advocates believe that laissez-faire individual-based recycling does not cover all of society's recycling needs. Thus, it does not negate the need for an organized recycling program. Local government can consider the activities of the recycling poor as contributing to property blight.

Public Participation Rates

Changes that have been demonstrated to increase recycling rates include:

- Single-stream recycling

- Pay as you throw fees for trash

"Between 1960 and 2000, the world production of plastic resins increased 25-fold, while recovery of the material remained below 5 percent." Many studies have addressed recycling behaviour and strategies to encourage community involvement in recycling programmes. It has been argued that recycling behaviour is not natural because it requires a focus and appreciation for long-term planning, whereas humans have evolved to be sensitive to short-term survival goals; and that to overcome this innate predisposition, the best solution would be to use social pressure to compel participation in recycling programmes. However, recent studies have concluded that social pressure is unviable in this context. One reason for this is that social pressure functions well in small group sizes of 50 to 150 individuals (common to nomadic hunter–gatherer peoples) but not in communities numbering in the millions, as we see today. Another reason is that individual recycling does not take place in the public view.

In a study done by social psychologist Shawn Burn, it was found that personal contact with individuals within a neighborhood is the most effective way to increase recycling within a community. In his study, he had 10 block leaders talk to their neighbors and persuade them to recycle. A comparison group was sent fliers promoting recycling. It was found that the neighbors that were personally contacted by their block leaders recycled much more than the group without personal contact. As a result of this study, Shawn Burn believes that personal contact within a small group of people is an important factor in encouraging recycling. Another study done by Stuart Oskamp examines the effect of neighbors and friends on recycling. It was found in his studies that people who had friends and neighbors that recycled were much more likely to also recycle than those who didn't have friends and neighbors that recycled.

Many schools have created recycling awareness clubs in order to give young students an insight on recycling. These schools believe that the clubs actually encourage students to not only recycle at school, but at home as well.

Plastic Recycling

DIVERTING PLASTICS FROM LANDFILLS: A TWO-PRONGED APPROACH

RECYCLING USED PLASTICS

A — ONCE DONE WITH PLASTIC

TURNING NON-RECYCLED PLASTICS INTO ENERGY

B

Plastic recycling is the process of recovering scrap or waste plastic and reprocessing the material into useful products. Since plastic is non-biodegradable, recycling it is a part of global efforts to reduce plastic in the waste stream, especially the approximately eight million metric tonnes of waste plastic that enter the Earth's ocean every year. This helps to reduce the high rates of plastic pollution.

Plastic recycling includes taking any type of plastic sorting it into different polymers and then chipping it and then melting it down into pellets after this stage it can then be used to make items of any kind such as plastic chairs and tables. Soft Plastics are also recycled such as polyethylene film and bags. This closed-loop operation has taken place since the 1970s and has made the production of some plastic products amongst the most efficient operations today.

Compared with lucrative recycling of metal, and similar to the low value of glass, plastic polymers recycling is often more challenging because of low density and low value. There are also numerous technical hurdles to overcome when recycling plastic.

A macro molecule interacts with its environment along its entire length, so total energy involved in mixing it is largely due to the product side stoichiometry. Heating alone is not enough to dissolve such a large molecule, so plastics must often be of nearly identical composition to mix efficiently.

When different types of plastics are melted together, they tend to phase-separate, like oil and water, and set in these layers. The phase boundaries cause structural weakness in the resulting material, meaning that polymer blends are useful in only limited applications.

Another barrier to recycling is the widespread use of dyes, fillers, and other additives in plastics. The polymer is generally too viscous to economically remove fillers, and would be damaged by many of the processes that could cheaply remove the added dyes. Additives are less widely used in beverage containers and plastic bags, allowing them to be recycled more often. Yet another barrier to removing large quantities of plastic from the waste stream and landfills is the fact that many common but small plastic items lack the universal triangle recycling symbol and accompanying number. An example is the billions of plastic utensils commonly distributed at fast food restaurants or sold for use at picnics.

The percentage of plastic that can be fully recycled, rather than downcycled or go to waste can be increased when manufacturers of packaged goods minimize mixing of packaging materials and eliminate contaminants. The Association of Plastics Recyclers have issued a Design Guide for Recyclability.

The use of biodegradable plastics is increasing.

Processes

Before recycling, most plastics are sorted according to their resin type. In the past, plastic reclaimers used the resin identification code (RIC), a method of categorization of polymer types, which was developed by the Society of the Plastics Industry in 1988. polyethylene terephthalate, commonly referred to as PET, for instance, has a resin code of 1. Most plastic reclaimers do not rely on the RIC now; they use automatic sort systems to identify the resin. Ranging from manual sorting and picking of plastic materials; to mechanized automation processes that involve shredding, sieving, separation by rates of density i.e. air, liquid, or magnetic, and complex spectrophotometric distribution technologies e.g. UV/VIS, NIR, Laser, etc. Some plastic products are also separated by color before they are recycled. The plastic recyclables are then shredded. These shredded fragments then undergo processes to eliminate impurities like paper labels. This material is melted and often extruded into the form of pellets which are then used to manufacture other products.

Thermal Depolymerization

Another process involves the conversion of assorted polymers into petroleum by a much less precise thermal depolymerization process. Such a process would be able to accept almost any polymer or mix of polymers, including thermoset materials such as vulcanized rubber tire separation of wastes and the biopolymers in feathers and other agricultural waste. Like natural petroleum, the chemicals produced can be made into fuels as well as polymers. A pilot plant of this type exists in Carthage, Missouri, United States, using turkey waste as input material. Gasification is a similar process, but is not technically recycling, since polymers are not likely to become the result.

Heat Compression

Yet another process that is gaining ground with startup companies (especially in Australia, United States and Japan) is heat compression. The heat compression process takes all unsorted, cleaned plastic in all forms, from soft plastic bags to hard industrial waste, and mixes the load in tumblers (large rotating drums resembling giant clothes dryers). The most obvious benefit to this method is the fact that all plastic is recyclable, not just matching forms. However, criticism rises from the energy costs of rotating the drums, and heating the post-melt pipes.

Distributed Recycling

For some waste plastics, recent technical devices called recyclebots enable a form of distributed recycling. Preliminary life-cycle analysis(LCA) indicates that such distributed recycling of HDPE to make filament of 3-D printers in rural regions is energetically favorable to either using virgin resin or conventional recycling processes because of reductions in transportation energy

Other Processes

A process has also been developed in which many kinds of plastic can be used as a carbon source in the recycling of scrap steel. There is also a possibility of mixed recycling of different plastics, which does not require their separation. It is called Compatibilization and requires use of special chemical bridging agents compatibilizers. It can help to keep the quality of recycled material and to skip often expensive and inefficient preliminary scanning of waste plastics streams and their separation/purification.

Applications

PET

Post-consumer polyethylene terephthalate (PET or PETE) containers are sorted into different colour fractions, and baled for onward sale. PET recyclers further sort the baled bottles and they are washed and flaked (or flaked and then washed). Non-PET fractions such as caps and labels are removed during this process. The clean flake is dried. Further treatment can take place e.g. melt filtering and pelletising or various treatments to produce food-contact-approved recycled PET (RPET).

RPET has been widely used to produce polyester fibres. This sorted post-consumer PET waste is crushed, chopped into flakes, pressed into bales, and offered for sale.

One use for this recycled PET that has recently started to become popular is to create fabrics to be used in the clothing industry. The fabrics are created by spinning the PET flakes into thread and yarn. This is done just as easily as creating polyester from brand new PET. The recycled PET thread or yarn can be used either alone or together with other fibers to create a very wide variety of fabrics. Traditionally these fabrics are used to create strong, durable, rough, products, such as jackets, coat, shoes, bags, hats, and accessories since they are usually too rough for direct skin contact and can cause irritation. However, these types of fabrics have become more popular as a result of the public's growing awareness of environmental issues. Numerous fabric and clothing manufacturers have capitalized on this trend.

Other major outlets for RPET are new containers (food-contact or non-food-contact) produced either by (injection stretch blow) moulding into bottles and jars or by thermoforming APET sheet to produce clam shells, blister packs and collation trays. These applications used 46% of all RPET produced in Europe in 2010. Other applications, such as strapping tape, injection-moulded engineering components and even building materials account for 13% of the 2010 RPET production.

In the United States the recycling rate for PET packaging was 31.2% in 2013, according to a report from The National Association for PET Container Resources (NAPCOR) and The Association of Postconsumer Plastic Recyclers (APR). A total of 1,798 million pounds was collected and 475 million pounds of recycled PET used out of a total of 5,764 million pounds of PET bottles.

HDPE

Plastic # 2, high-density polyethylene (HDPE) is a commonly recycled plastic. It is typically downcycled into plastic lumber, tables, roadside curbs, benches, truck cargo liners, trash receptacles, stationery (e.g. rulers) and other durable plastic products and is usually in demand.

PS

Most polystyrene products are currently not recycled due to the lack of incentive to invest in the compactors and logistical systems required. As a result, manufacturers cannot obtain sufficient scrap. Expanded polystyrene (EPS) scrap can easily be added to products such as EPS insulation sheets and other EPS materials for construction applications. When it is not used to make more EPS, foam scrap can be turned into clothes hangers, park benches, flower pots, toys, rulers, stapler bodies, seedling containers, picture frames, and architectural molding from recycled PS.

The resin identification code symbol for polystyrene

Recycled EPS is also used in many metal casting operations. Rastra is made from EPS that is combined with cement to be used as an insulating amendment in the making of concrete foundations and walls. Since 1993, American manufacturers have produced insulating concrete forms made with approximately 80% recycled EPS.

Other Plastics

The white plastic polystyrene foam peanuts used as packing material are often accepted by shipping stores for reuse.

Successful trials in Israel have shown that plastic films recovered from mixed municipal waste streams can be recycled into useful household products such as buckets.

Similarly, agricultural plastics such as mulch film, drip tape and silage bags are being diverted from the waste stream and successfully recycled into much larger products for industrial applications such as plastic composite railroad ties. Historically, these agricultural plastics have primarily been either landfilled or burned on-site in the fields of individual farms.

CNN reports that Dr. S. Madhu of the Kerala Highway Research Institute, India, has formulated a road surface that includes recycled plastic: aggregate, bitumen (asphalt) with plastic that has been shredded and melted at a temperature below 220 degrees C (428 °F) to avoid pollution. This road surface is claimed to be very durable and monsoon rain resistant. The plastic is sorted by hand, which is economical in India. The test road used 60 kg of plastic for an approximately 500m-long, 8m-wide, two-lane road. The process chops thin-film road-waste into a light fluff of tiny flakes that hot-mix plants can uniformly introduce into viscous bitumen with a customized dosing machine. Tests at both Bangalore and the Indian Road Research Centre indicate that roads built using this 'KK process' will have longer useful lives and better resistance to cold, heat, cracking, and rutting, by a factor of three.

Recycling Rates

The quantity of post-consumer plastics recycled has increased every year since at least 1990, but rates lag far behind those of other items, such as newspaper (about 80%) and corrugated fiberboard (about 70%). Overall, U.S. post-consumer plastic waste for 2008 was estimated at 33.6 million tons; 2.2 million tons (6.5%) were recycled and 2.6 million tons (7.7%) were burned for energy; 28.9 million tons, or 85.5%, were discarded in landfills.

Economic and Energy Potential

In 2008, the price of PET dropped from $370/ton in the US to $20 in November. PET prices had returned to their long-term averages by May 2009.

Recycling one ton of plastic can save 5,774 kWh of energy, 98,000,000 btus of energy, 1,000-2,000 gallons of gasoline, 685 gallons of oil, 30 cubic yards of landfill space, 48,000 gallons of water.

Consumer Education

United Kingdom

In the UK, the amount of post-consumer plastic being recycled is relatively low, due in part to a lack of recycling facilities.

The Plastics 2020 Challenge was founded in 2009 by the plastics industry with the aim of engaging the British public in a nationwide debate about the use, reuse and disposal of plastics, and hosts a series of online debates on its website framed around the waste hierarchy.

There is a facility in Worksop capable of recycling 60–80 thousand metric tonnes a year.

In Northern Ireland, the rate of recycling is relatively low at only 37.4%. However, emerging technologies are helping to increase the recycling rates of items previously landfilled e.g. mixed hard plastics.

Plastic Identification Code

Five groups of plastic polymers, each with specific properties, are used worldwide for packaging applications. Each group of plastic polymer can be identi-fied by its Plastic Identification code (PIC), usually a number or a letter abbreviation. For instance, Low-Density Polyethylene can be identified by the number "4" or the letters "LDPE". The PIC appears inside a three-chasing-arrow recycling symbol. The symbol is used to indicate whether the plastic can be recycled into new products.

The PIC was introduced by the Society of the Plastics Industry, Inc., to provide a uni-form system for the identification of various polymer types and to help recycling companies separate various plastics for reprocessing. Manufacturers of plastic products are required to use PIC labels in some countries/regions and can voluntarily mark their products with the PIC where there are no requirements. Consumers can identify the plastic types based on the codes usually found at the base or at the side of the plastic products, including food/chemical packaging and containers. The PIC is usually not present on packaging films, since it is not practical to collect and recycle most of this type of waste.

United States

Low national plastic recycling rates have been due to the complexity of sorting and processing, unfavorable economics, and consumer confusion about which plastics can actually be recycled. Part of the confusion has been due to the use of the resin identification code which is not on all plastic parts but just a subset that includes the recycling symbol as part of its design. The resin identification code is stamped or printed on the bottom of containers and surrounded by a triangle of arrows. The intent of these symbols was to make it easier to identify the type of plastics used to make a particular container and to indicate that the plastic is potentially recyclable. The question that remains is which types of plastics can be recycled by your local recycling center. In many communities, not all types of plastics are accepted for sidewalk recycling collection programs due to the high processing costs and complexity of the equipment required to recycle certain materials. There is also sometimes a seemingly low demand for the recycled product depending on a recycling center's proximity to entities seeking recycled materials. Another major barrier is that the cost to recycle certain materials and the corresponding market price for those materials sometimes does not present any opportunity for profit. The best example of this is polystyrene (commonly called styrofoam), although some communities, like Brookline, MA, are moving toward banning the distribution of polystyrene containers by local food and coffee businesses.

Computer Recycling

Computer recycling, electronic recycling or e-waste recycling is the disassembly and separation of components and raw materials of waste electronics. Although the procedures of re-use, donation and repair are not strictly recycling, they are other common sustainable ways to dispose of IT waste.

Computer monitors are typically packed into low stacks on wooden pallets for recycling and then shrink-wrapped.

In 2009, 38% of computers and a quarter of total electronic waste was recycled in the United States, 5% and 3% up from 3 years prior respectively. Since its inception in the early 1990s, more and more devices are recycled worldwide due to increased awareness and investment. Electronic recycling occurs primarily in order to recover valuable rare earth metals and precious metals, which are in short supply, as well as plastics and metals. These are resold or used in new devices after purification, in effect creating a circular economy.

Recycling is considered environmentally friendly because it prevents hazardous waste, including heavy metals and carcinogens, from entering the atmosphere, landfill or waterways. While electronics consist a small fraction of total waste generated, they are far more dangerous. There is stringent legislation designed to enforce and encourage the sustainable disposal of appliances, the most notable being the Waste Electrical and Electronic Equipment Directive of the European Union and the United States National Computer Recycling Act.

Opponents argue that recycling is expensive and ineffective, that it does not safeguard data and that it stifles innovation. It is also criticised for exporting, often illegally, large volumes of toxic waste to countries such as India, China and Nigeria for crude manual disassembly by workers who have little regard for the risk to themselves or the environment.

Reasons for Recycling

Obsolete computers and old electronics are valuable sources for secondary raw materials if recycled; otherwise, these devices are a source of toxins and carcinogens. Rapid technology change, low initial cost, and planned obsolescence have resulted in a fast-growing surplus of computers and other electronic components around the globe. Technical solutions are available, but in most cases a legal framework, collection system, logistics, and other services need to be implemented before applying a technical solution. The U.S. Environmental Protection Agency, estimates 30 to 40 million surplus PCs, classified as "hazardous household waste", would be ready for end-of-life management in the next few years. The U.S. National Safety Council estimates that 75% of all personal computers ever sold are now surplus electronics.

In 2007, the United States Environmental Protection Agency (EPA) stated that more than 63 million computers in the U.S. were traded in for replacements or discarded. Today, 15% of electronic devices and equipment are recycled in the United States. Most electronic waste is sent to landfills or incinerated, which releases materials such as lead, mercury, or cadmium into the soil, groundwater, and atmosphere, thus having a negative impact on the environment.

Many materials used in computer hardware can be recovered by recycling for use in future production. Reuse of tin, silicon, iron, aluminium, and a variety of plastics that are present in bulk in computers or other electronics can reduce the costs of constructing new systems. Components frequently contain lead, copper, gold and other valuable materials suitable for reclamation.

Computer components contain many toxic substances, like dioxins, polychlorinated biphenyls (PCBs), cadmium, chromium, radioactive isotopes and mercury. A typical computer monitor may contain more than 6% lead by weight, much of which is in the lead glass of the cathode ray tube (CRT). A typical 15 inch (38 cm) computer monitor may contain 1.5 pounds (1 kg) of lead but other monitors have been estimated to have up to 8 pounds (4 kg) of lead. Circuit boards contain considerable quantities of lead-tin solders that are more likely to leach into groundwater or create air pollution due to incineration. The processing (e.g. incineration and acid treatments) required to reclaim these precious substances may release, generate, or synthesize toxic by-products.

Export of waste to countries with lower environmental standards is a major concern. The Basel Convention includes hazardous wastes such as, but not limited to, CRT screens as an item that may not be exported transcontinentally without prior consent of both the country exporting and receiving the waste. Companies may find it cost-effective in the short term to sell outdated computers to less developed countries with lax regulations. It is commonly believed that a majority of surplus laptops are routed to developing nations as "dumping grounds for e-waste". The high value of working and reusable laptops, computers, and components (e.g. RAM) can help pay the cost of transportation for many worthless "commodities". The laws governing the exportation of waste electronics are put in place to stop "recycling companies" in developed countries from shipping their waste to 3rd world countries as working devices; they are never working devices. The 3rd world workers scavenge specific items with selling value and throw the rest away to rot and become a health hazard in their own backyard.

Regulations

In Switzerland, the first electronic waste recycling system was implemented in 1991, beginning with collection of old refrigerators; over the years, all other electric and electronic devices were gradually added to the system. The established producer responsibility organization is SWICO, mainly handling information, communication, and organization technology. The European Union implemented a similar system in February 2003, under the Waste Electrical and Electronic Equipment Directive (WEEE Directive, 2002/96/EC).

Pan European adoption of the Legislation was slow on take-up, with Italy and the United Kingdom being the final member states to pass it into law. The success of the WEEE directive has varied significantly from state to state, with collection rates varying between 13 kilograms per capita per annum to as little as 1 kg per capita per annum. Computers & electronic wastes collected from households within Europe are treated under the WEEE directive via Producer Compliance Schemes (whereby manufacturers of Electronics pay into a scheme that funds its recovery from household waste recycling centres (HWRCs)) and nominated Waste Treatment Facilities (known as Obligated WEEE).

An abandoned Taxan monitor.

Europe

However, recycling of ex corporate Computer Hardware and associated electronic equipment falls outside the Producer Compliance Scheme (Known as non-obligated). In the UK, Waste or obsolete corporate related computer hardware is treated via third party Authorized Treatment Facilities, who normally impose a charge for its collection and treatment.

United States

Federal

The United States Congress considers a number of electronic waste bills, like the National Computer Recycling Act introduced by Congressman Mike Thompson (D-CA). The main federal law governing solid waste is the Resource Conservation and Recovery Act of 1976. It covers only CRTs, though state regulations may differ. There are also separate laws concerning battery disposal. On March 25, 2009, the House Science and Technology Committee approved funding for research on reducing electronic waste and mitigating environmental impact, regarded by sponsor Ralph Hall (R-TX) as the first federal bill to directly address electronic waste.

State

Many states have introduced legislation concerning recycling and reuse of computers or computer parts or other electronics. Most American computer recycling legislations address it from within the larger electronic waste issue.

In 2001, Arkansas enacted the Arkansas Computer and Electronic Solid Waste Management Act, which requires that state agencies manage and sell surplus computer equipment, establishes a computer and electronics recycling fund, and authorizes the Department of Environmental Quality to regulate and/or ban the disposal of computer and electronic equipment in Arkansas landfills.

The recently passed Electronic Device Recycling Research and Development Act distributes grants to universities, government labs and private industries for research in developing projects in line with e-waste recycling and refurbishment.

Asia

In Japan, sellers and manufacturers of certain electronics (such as televisions and air conditioners) are required to recycle them. However, no legislation exists to cover the recycling of computer or cellphone related wastes.

It is required in South Korea and Taiwan that sellers and manufacturers of electronics be responsible for recycling 75% of their used products.

According to a report by UNEP titled, "Recycling - from E-Waste to Resources," the amount of e-waste being produced - including mobile phones and computers - could rise by as much as 500 percent over the next decade in some countries, such as India.

Electronic waste is often exported to developing countries.

4.5-volt, D, C, AA, AAA, AAAA, A23, 9-volt, CR2032 and LR44 cells are all recyclable in most countries.

One theory is that increased regulation of electronic waste and concern over the environmental harm in mature economies creates an economic disincentive to remove residues prior to export. Critics of trade in used electronics maintain that it is too easy for brokers calling themselves recyclers to export unscreened electronic waste to developing countries, such as China, India and parts of Africa, thus avoiding the expense of removing items like bad cathode ray tubes (the processing of which is expensive and difficult). The developing countries are becoming big dump yards of e-waste. Proponents of international trade point to the success of fair trade programs in other industries, where cooperation has led creation of sustainable jobs, and can bring affordable technology in countries where repair and reuse rates are higher.

Organizations like A2Z Group (Company Website) have stepped in to own up the responsibility to collect and recycle e-waste at various locations in India.

South Africa

Thanks to the National Environmental Management Act 1998 and National Environmental Management Waste Act 2008, any person in any position causing harm to the environment and failing to comply with the Waste Act could be fined R10 Million or put into jail or receive both penalties for their transgressions.

Recycling Methods

Computers being collected for recycling at a pickup event in Olympia, Washington, United States.

Consumer Recycling

Consumer recycling options consists of sale, donating computers directly to organizations in need, sending devices directly back to their original manufacturers, or getting components to a convenient recycler or refurbisher.

Scrapping/Recycling

The rising price of precious metals — coupled with the high rate of unemployment during the Great Recession — has led to a larger number of amateur "for profit" electronics recyclers. Computer parts, for example, are stripped of their most valuable components and sold for scrap. Metals like copper, aluminum, lead, gold and palladium are recovered from computers, televisions and more.

In the recycling process, TVs, monitors, mobile phones and computers are typically tested for reuse and repaired. If broken, they may be disassembled for parts still having high value if labour is cheap enough. Other e-waste is shredded to roughly 100 mm pieces and manually checked to separate out toxic batteries and capacitors which contain poisonous metals. The remaining pieces are further shredded to ~10 mm and passed under a magnet to remove ferrous metals. An eddy current ejects non-ferrous metals, which are sorted by density either by a centrifuge or vibrating plates. Precious metals can be dissolved in acid, sorted, and smelted into ingots. The remaining glass and plastic fractions are separated by density and sold to re-processors. TVs and monitors must be manually disassembled to remove either toxic lead in CRTs or the mercury in flat screens.

Bulk laptops at a recycling affiliate, broken down into Dell, Gateway Computers, Hewlett-Packard, Sony, and other.

Corporate Recycling

Businesses seeking a cost-effective way to recycle large amounts of computer equipment responsibly face a more complicated process.

Businesses also have the options of sale or contacting the Original Equipment Manufacturers (OEMs) and arranging recycling options.

Some companies pick up unwanted equipment from businesses, wipe the data clean from the systems, and provide an estimate of the product's remaining value. For unwanted items that still have value, these firms buy the excess IT hardware and sell refurbished products to those seeking more affordable options than buying new.

Companies that specialize in data protection and green disposal processes dispose of both data and used equipment, while employing strict procedures to help improve the environment. Professional IT Asset Disposition (ITAD) firms specialize in corporate computer disposal and recycling services in compliance with local laws and regulations and also offer secure data elimination services that comply with Data remanence standards including National Institute of Standards and Technology.

Corporations face risks both for incompletely destroyed data and for improperly disposed computers. In the UK, some recycling companies use a specialized WEEE-registered contractor to dispose IT equipment and electrical appliances, who disposes it safely and legally. In America, companies are liable for compliance with regulations even if the recycling process is outsourced under the Resource Conservation and Recovery Act. Companies can mitigate these risks by requiring waivers of liability, audit trails, certificates of data destruction, signed confidentiality agreements, and random audits of information security. The National Association of Information Destruction is an international trade association for data destruction providers.

Sale

Online auctions are an alternative for consumers willing to resell for cash less fees, in a complicated, self-managed, competitive environment where paid listings might not sell. Online classified ads can be similarly risky due to forgery scams and uncertainty.

Take Back

When researching computer companies before a computer purchase, consumers can find out if they offer recycling services. Most major computer manufacturers offer some form of recycling. At the user's request they may mail in their old computers, or arrange for pickup from the manufacturer.

Hewlett-Packard also offers free recycling, but only one of its "national" recycling programs is available nationally, rather than in one or two specific states. Hewlett-Packard also offers to pick up any computer product of any brand for a fee, and to offer a coupon against the purchase of future computers or components; it was the largest computer recycler in America in 2003, and it has recycled over 750,000,000 pounds (340,000,000 kg) of electronic waste globally since 1995. It encourages the shared approach of collection points for consumers and recyclers to meet.

Exchange

Manufacturers often offer a free replacement service when purchasing a new PC. Dell Computers and Apple Inc. take back old products when one buys a new one. Both refurbish and resell their own computers with a one-year warranty.

Many companies purchase and recycle all brands of working and broken laptops and notebook computers from individuals and corporations. Building a market for recycling of desktop computers has proven more difficult than exchange programs for laptops, smartphones and other smaller electronics. A basic business model is to provide a seller an instant online quote based on laptop characteristics, then to send a shipping label and prepaid box to the seller, to erase, reformat, and process the laptop, and to pay rapidly by cheque. A majority of these companies are also generalized electronic waste recyclers as well; organizations that recycle computers exclusively include Cash For Laptops, a laptop refurbisher in Nevada that claims to be the first to buy laptops online, in 2001.

Donations/Nonprofits

With the constant rising costs due to inflation, many families or schools do not have the sufficient funds available for computers to be utilized along with education standards. Families also impacted by disaster suffer as well due to the financial impact of the situation they have incurred. Many nonprofit organizations, such as InterConnection.org, can be found locally as well as around the web and give detailed descriptions as to what methods are used for dissemination and detailed instructions on how to donate. The impact can be seen locally and globally, affecting thousands of those in need. In Canada non profit organizations engaged in computer recycling, such as The Electronic Recycling Association Calgary, Edmonton, Vancouver, Winnipeg, Toronto, Montreal, Computers for Schools Canada wide, are very active in collecting and refurbishing computers and laptops to help the non profit and charitable sectors and schools.

Junkyard Computing

The term *junkyard computing* is a colloquial expression for using old or inferior hardware to fulfill computational tasks while handling reliability and availability on software level. It utilizes abstraction of computational resources via software, allowing hardware replacement at very low effort. Ease of replacement is hereby a corner point since hardware failures are expected at any time due to the condition of the underlying infrastructure. This paradigm became more widely used with the introduction of cluster orchestration software like Kubernetes or Apache Mesos, since large monolithic applications require reliability and availability on machine level whereas this kind of software is fault tolerant by design. Those orchestration tools also introduced fairly fast set-up processes allowing to use junkyard computing economically and even making this pattern applicable in the first place. Further use cases were introduced when continuous delivery was getting more widely accepted. Infrastructure to execute tests and static code analysis was needed which requires as much performance as possible while being extremely cost effective. From an economical and technological perspective, junkyard computing is only practicable for a small amount of users or companies. It already requires a descend amount of physical machines to compensate hardware failures while maintaining the required reliability and availability. This implies a direct need for a matching underling infrastructure to house all the computers and servers. Scaling this paradigm is also quiet limited due to the increasing importance of factors like power efficiency and maintenance efforts, making this kind of computing perfect for mid-sized applications.

History

Although consumer electronics such as the radio have been popular since the 1920s, recycling was almost unheard of until the early 1990s. At the end of the 1970s the accelerating pace of domestic consumer electronics drastically shortened the lifespan of electronics such as TVs, VCRs and audio. New innovations appeared more quickly, making older equipment considered obsolete. Increased complexity and sophistication of manufacture made local repair more difficult. The retail market shifted gradually, but substantially from a few high-value items that were cherished for years and repaired when necessary, to short-lived items that were rapidly replaced owing to wear or simply fashion, and discarded rather than repaired. This was particularly evident in computing, highlighted by Moore's Law. In 1988 two severe incidents highlighted the approaching e-waste crisis. The cargo barge Khian Sea, was loaded with more than 14,000 tons of toxic ash from Pennsylvania which had been refused acceptance in New Jersey and the

Caribbean. After sailing for 16 months, all the waste was dumped as "topsoil fertiliser" in Haiti and in the Bay of Bengal by November 1988. In June 1988, a large illegal toxic waste dump which had been created by an Italian company was discovered. This led to the formation of the Basel Convention to stem the flow of poisonous substances from developed countries in 1989.

In 1991, the first electronic waste recycling system was implemented in Switzerland, beginning with collection of old refrigerators but gradually expanding to cover all devices. The organisation SWICO handles the programme, and is a partnership between IT retailers.

The first publication to report the recycling of computers and electronic waste was published on the front page of the New York Times on April 14, 1993 by columnist Steve Lohr. It detailed the work of Advanced Recovery Inc., a small recycler, in trying to safely dismantle computers, even if most waste was landfilled. Several other companies emerged in the early 1990s, chiefly in Europe, where national 'take back' laws compelled retailers to use them.

After these schemes were set up, many countries did not have the capacity to deal with the sheer quantity of e-waste they generated or its hazardous nature. They began to export the problem to developing countries without enforced environmental legislation. This is cheaper: the cost of recycling of computer monitors in the US is ten times more than in China. Demand in Asia for electronic waste began to grow when scrap yards found they could extract valuable substances such as copper, iron, silicon, nickel and gold, during the recycling process.

The Waste Electrical and Electronic Equipment Directive (WEEE Directive) became European Law in February 2003 and covers all aspects of recycling all types of appliance. This was followed by Electronic Waste Recycling Act, enshrined in Californian law in January 2005

The 2000s saw a large increase in both the sale of electronic devices and their growth as a waste stream: in 2002 e-waste grew faster than any other type of waste in the EU. This caused investment in modern, automated facilities to cope with the influx of redundant appliances.

E-cycling

"E-cycling" or "E-waste" is an initiative by the United States Environmental Protection Agency (EPA) which refers to donations, reuse, shredding and general collection of used electronics. Generically, the term refers to the process of collecting, brokering, disassembling, repairing and recycling the components or metals contained in used or discarded electronic equipment, otherwise known as electronic waste (e-waste). "E-cyclable" items include, but are not limited to: televisions, computers, microwave ovens, vacuum cleaners, telephones and cellular phones, stereos, and VCRs and DVDs just about anything that has a cord, light or takes some kind of battery.

Investment in e-cycling facilities has been increasing recently due to technology's rapid rate of obsolescence, concern over improper methods, and opportunities for manufacturers to influence the secondary market (used and reused products). The higher metal prices is also having more recycling taking place. The controversy around methods stems from a lack of agreement over preferred outcomes.

World markets with lower disposable incomes, consider 75% repair and reuse to be valuable enough to justify 25% disposal. Debate and certification standards may be leading to better definitions, though civil law contracts, governing the expected process are still vital to any contracted process, as poorly defined as "e-cycling".

Pros of E-cycling

The e-waste disposal occurring after processing for reuse, repair of equipment, and recovery of metals may be unethical or illegal when e-scrap of many kinds is transported overseas to developing countries for such processing. It is transported as if to be repaired and/or recycled, but after processing the less valuable e-scrap becomes e-waste/pollution there. Another point of view is that the net environmental cost must be compared to and include the mining, refining and extraction with its waste and pollution cost of new products manufactured to replace secondary products which are routinely destroyed in wealthier nations, and which cannot economically be repaired in older or obsolete products. As an example of negative impacts of e-waste, pollution of groundwater has become so serious in areas surrounding China's landfills that water must be shipped in from 18 miles (29 km) away. However, mining of new metals can have even broader impacts on groundwater. Either thorough e-cycling processing, domestic processing or overseas repair, can help the environment by avoiding pollution. Such e-cycling can theoretically be a sustainable alternative to disposing of e-waste in landfills. In addition, e-cycling allows for the reclamation of potential conflict minerals, like gold and wolframite, which requires less of those to be mined and lessens the potential money flow to militias and other exploitative actors in third-world that profit from mining them.

Supporters of one form of "required e-cycling" legislation argue that e-cycling saves taxpayers money, as the financial responsibility would be shifted from the taxpayer to the manufacturers. Advocates of more simple legislation (such as landfill bans for e-waste) argue that involving manufacturers does not reduce the cost to consumers, if reuse value is lost, and the resulting costs are then passed on to consumers in new products, particularly affecting markets which can hardly afford new products. It is theorized that manufacturers who take part in e-cycling would be motivated to use fewer materials in the production process, create longer lasting products, and implement safer, more efficient recycling systems. This theory is sharply disputed and has never been demonstrated.

Criticisms of E-cycling

The critics of e-cycling are just as vocal as its advocates. According to the Reason Foundation, e-cycling only raises the product and waste management costs of e-waste for consumers and limits innovation on the part of high-tech companies. They also believe that e-cycling facilities could unintentionally cause great harm to the environment. Critics claim that e-waste doesn't occupy a significant portion of total waste. According to a European study, only 4% of waste is electronic.

Another opposition to e-cycling is that many problems are posed in disassembly: the process is costly and dangerous because of the heavy metals of which the electronic products are composed, and as little as 1-5% of the original cost of materials can be retrieved. A final problem that people find is that identity fraud is all too common in regards to the disposal of electronic products. As the programs are legislated, creating winners and losers among e-cyclers with different locations and processes, it may be difficult to distinguish between criticism of e-cycling as a practice, and criticism of the specific legislated means proposed to enhance it.

The Fate of E-waste

A hefty criticism often lobbed at reuse based recyclers is that people think that they are recycling their electronic waste, when in reality it is actually being exported to developing countries like China, India, and Nigeria. For instance, at free recycling drives, "recyclers" may not be staying true to their word, but selling e-waste overseas or to parts brokers. Studies indicate that 50-80% of the 300,000 to 400,000 tons (270,000 to 360,000 tonnes) of e-waste is being sent overseas, and that approximately 2 million tons (1.8 million tonnes) per year go to U.S. landfills.

Although not possible in all circumstances, the best way to e-cycle is to upcycle e-waste. On the other hand, the electronic products in question are generally manufactured, and repaired under warranty, in the same nations, which anti-reuse recyclers depict as primitive. Reuse-based e-recyclers believe that fair-trade incentives for export markets will lead to better results than domestic shredding. There has been a continued debate between export-friendly e-cycling and increased regulation of that practice.

In the European Union, debate regarding the export of e-waste has resulted in a significant amendment to the WEEE directive (January 2012) with a view to significantly diminishing the export of WEEE (untreated e-waste). During debate in Strasburg, MEPs stated that "53 million tonnes of WEEE were generated in 2009 but only 18% collected for recycling" with the remainder being exported or sent to landfill. The Amendment, voted through by a unanimous 95% of representatives, removed the re-use (repair and refurbishmet) aspect of the directive and placed more emphasis upon recycling and recovery of precious metals and base metals. The changes went further by placing the burden upon registered exporters to prove that used equipment leaving Europe was "fit for purpose".

Policy Issues and Current Efforts

Currently, pieces of government legislation and a number of grassroots efforts have contributed to the growth of e-cycling processes which emphasize decreased exports over increased reuse rates. The Electronic Waste Recycling Act was passed in California in 2003. It requires that consumers pay an extra fee for certain types of electronics, and the collected money be then redistributed to recycling companies that are qualified to properly recycle these products. It is the only state that legislates against e-waste through this kind of consumer fee; the other states' efforts focus on producer responsibility laws or waste disposal bans. No study has shown that per capita recovery is greater in one type of legislated program (e.g. California) versus ordinary waste disposal bans (e.g. Massachusetts), though recovery has greatly increased in states which use either method.

As of September, 2006, Dell developed the nation's first completely free recycling program, furthering the responsibilities that manufacturers are taking for e-cycling. Manufacturers and retailers such as Best Buy, Sony, and Samsung have also set up recycling programs. This program does not accept televisions, which are the most expensive used electronic item, and are unpopular in markets which must deal with televisions when the more valuable computers have been cherry picked.

Another step being taken is the recyclers' pledge of true stewardship, sponsored by the Computer TakeBack Campaign. It has been signed by numerous recyclers promising to recycle responsibly. Grassroots efforts have also played a big part in this issue, as they and other community organizations are being formed to help responsibly recycle e-waste. Other grassroots campaigns are Basel, the Computer TakeBack Campaign (co-coordinated by the Grassroots Recycling Network), and the Silicon Valley Toxics Coalition. No study has shown any difference in recycling methods under the Pledge, and no data is available to demonstrate difference in management between "Pledge" and non-Pledge companies, though it is assumed that the risk of making false claims will prevent Pledge companies from wrongly describing their processes.

Many people believe that the U.S. should follow the European Union model in regards to its management of e-waste. In this program, a directive forces manufacturers to take responsibility for e-cycling; it also demands manufacturers' mandatory take-back and places bans on exporting e-waste to developing countries. Another longer-term solution is for computers to be composed of less dangerous products and many people disagree. No data has been provided to show that people who agree with the European model have based their agreement on measured outcomes or experience-based scientific method.

Data Security

E-waste presents a potential security threat to individuals and exporting countries. Hard drives that are not properly erased before the computer is disposed of can be reopened, exposing sensitive information. Credit card numbers, private financial data, account information and records of online transactions can be accessed by most willing individuals. Organized criminals in Ghana commonly search the drives for information to use in local scams.

Electronic waste dump at Agbogbloshie, Ghana. Organized criminals commonly search the drives for information to use in local scams.

Government contracts have been discovered on hard drives found in Agbogbloshie, Ghana. Multimillion-dollar agreements from United States security institutions such as the Defense Intelligence Agency (DIA), the Transportation Security Administration and Homeland Security have all resurfaced in Agbogbloshie.

Reasons to Destroy and Recycle Securely

There are ways to ensure that not only hardware is destroyed but also the private data on the hard drive. Having customer data stolen, lost, or misplaced contributes to the ever growing number of people who are affected by identity theft, which can cause corporations to lose more than just money. The image of a company that holds secure data, such as banks, law firms, pharmaceuticals, and credit corporations is also at risk. If a company's public image is hurt, it could cause consumers to not use their services and could cost millions in business losses and positive public relation campaigns. The cost of data breaches "varies widely, ranging from $90 to $50,000 (under HIPAA's new HITECH amendment, that came about through the American Recovery and Revitalization act of 2009), as per customer record, depending on whether the breach is "low-profile" or "high-profile" and the company is in a non-regulated or highly regulated area, such as banking or medical institutions."

There is also a major backlash from the consumer if there is a data breach in a company that is supposed to be trusted to protect their private information. If an organization has any consumer info on file, they must by law (Red Flags Clarification act of 2010) have written information protection policies and procedures in place, that serve to combat, mitigate, and detect vulnerable areas that could result in identity theft. The United States Department of Defense has published a standard to which recyclers and individuals may meet in order to satisfy HIPAA requirements.

Secure Recycling

Countries have developed standards, aimed at businesses and with the purpose of ensuring the security of Data contained in 'confidential' computer media [NIST 800-88: US standard for Data Remenance][HMG CESG IS5, Baseline & Enhanced, UK Government Protocol for Data Destruction]. National Association for Information Destruction (NAID) "is the international trade association for companies providing information destruction services. Suppliers of products, equipment and services to destruction companies are also eligible for membership. NAID's mission is to promote the information destruction industry and the standards and ethics of its member companies." There are companies that follow the guidelines from NAID and also meet all Federal EPA and local DEP regulations.

The typical process for computer recycling aims to securely destroy hard drives while still recycling the byproduct. A typical process for effective computer recycling:

1. Receive hardware for destruction in locked and securely transported vehicles.

2. Shred hard drives.

3. Separate all aluminum from the waste metals with an electromagnet.

4. Collect and securely deliver the shredded remains to an aluminum recycling plant.

5. Mold the remaining hard drive parts into aluminum ingots.

The Asset Disposal and Information Security Alliance (ADISA) publishes an *ADISA IT Asset Disposal Security Standard* that covers all phases of the e-waste disposal process from collection to transportation, storage and sanitization's at the disposal facility. It also conducts periodic audits of disposal vendors.

Ship Breaking

Ship breaking or ship demolition is a type of ship disposal involving the breaking up of ships for either a source of parts, which can be sold for re-use, or for the extraction of raw materials, chiefly scrap. It may also be known as ship dismantling, ship cracking, or ship recycling. Modern ships have a lifespan of 25 to 30 years before corrosion, metal fatigue and a lack of parts render them uneconomical to run. Ship breaking allows the materials from the ship, especially steel, to be recycled and made into new products.

This lowers the demand for mined iron ore and reduces energy use in the steel-making process. Equipment on board the vessel can also be reused. While ship breaking is, in theory, sustainable, there are concerns about the use of poorer countries without stringent environmental legislation. It is also considered one of the world's most dangerous industries and very labour-intensive.

Workers drag steel plate ashore from beached ships in Chittagong, Bangladesh

In 2012, roughly 1,250 ocean ships were broken down, and their average age was 26 years. In 2013, the world total of demolished ships amounted to 29,052,000 tonnes, 92% of which were demolished in Asia. India, Bangladesh, China and Pakistan have the highest market share and are global centres of ship breaking, with Alang in India and Gadani in Pakistan being the largest ships graveyards in the world. The largest sources of ships are states of China, Greece and Germany respectively, although there is a greater variation in the source of carriers versus their disposal. The ship breaking yards of India, Bangladesh, China and Pakistan employ 100,000 workers as well as providing a large amount of indirect jobs. In India, the recycled steel covers 10% of the country's needs.

As an alternative to ship breaking, ships may be sunk to create artificial reefs after being cleared of hazardous materials, or sunk in deep ocean waters. Storage is a viable temporary option, whether on land or afloat, though all ships will be eventually scrapped, sunk, or preserved for museums.

History

Wooden-walled ships were simply set on fire or 'conveniently sunk'. In Tudor times, ships were also dismantled and the timber re-used. This procedure was no longer applicable with the advent of metal-hulled boats.

There is evidence of the industry as early as 1838. The navy vessel HMS Temeraire had her masts, stores and guns removed and her crew paid off. She was sold by Dutch auction on 16 August 1838 to John Beatson, a shipbreaker based at Rotherhithe for £5,530. Beatson was then faced with the task of transporting the ship 55 miles from Sheerness to Rotherhithe, the largest ship to have attempted this voyage. To accomplish this he hired two steam tugs from the Thames Steam Towing Company and employed a Rotherhithe pilot named William Scott and twenty five men to sail her up the Thames, at a cost of £58. The shipbreakers undertook a thorough dismantling, removing all the copper sheathing, rudder pintles and gudgeons, copper bolts, nails and other fastenings to be sold back to the Admiralty. The timber was mostly sold to house builders and shipyard owners, though some was retained for working into specialist commemorative furniture. The ship's final voyage was immortalised by William Turner's painting The Fighting 'Temeraire', Tugged to her Last Berth to be Broken Up, 1838.

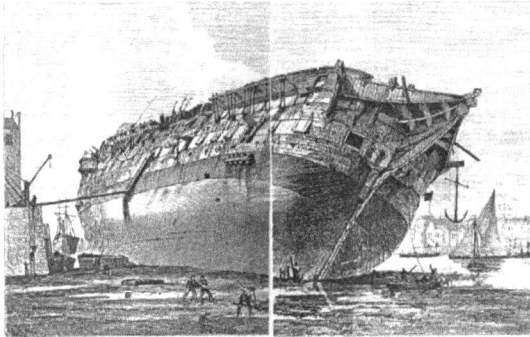

HMS *Queen* heeled over on the Thames foreshore off Rotherhithe.

In 1880, Denny Brothers of Dumbarton used scrap maritime steel in their shipbuilding. Many other nations began to purchase British ships for scrap by the late 19th century, including Germany, Italy, the Netherlands and Japan. The Italian industry started in 1892, and the Japanese after an 1896 law had been passed to subsidise native ship-building.

After being damaged or involved in a disaster, liner operators did not want the name of the broken ship to tarnish the brand of their passenger services. The final voyage of many Victorian ships was with the final letter of their name chipped off.

The armistice of 1918 brought a glut of naval 'men-o-war' into the ship breaking industry which boomed and subsequently died down in the early 1920s. In the 1930s, it became cheaper to 'beach' a boat and run her ashore as opposed to using a dry dock. The ship would have to weigh as little as possible and run ashore at full speed. Dismantling operations required a 10 feet rise of tide and close proximity to a steel-works. Electric shears, a wrecking ball and oxy-acetylene torches were used. The technique of the time is almost identical to that of developing countries today. Similarly, Thos W Ward Ltd., one of the largest breakers in the United Kingdom in the 1930s, would recondition and sell all furniture and machinery. Many historical artefacts were sold at public auctions: the Cunarder Mauretania received high bids for her fittings worldwide. However, even with obsolete technology, any weapons and military information were carefully removed.

Until the late 20th century, ship breaking took place in port cities of industrialized countries such as the United Kingdom and the United States. Those dismantlers that still remain in the United States work primarily on government surplus vessels. In the mid 20th century, low-cost East Asian countries began to dominate ship breaking, with countries such as Japan, then Korea and Taiwan and then China increasing their world share. For example, in 1977 Taiwan dominated the industry with more than half the market share, followed by Spain and Pakistan. Bangladesh had no capacity at all. However, the sector is volatile and fluctuates wildly, and Taiwan processed just 2 ships 13 years later as wages across East Asia rose.

Dismantling of *Redoutable* in Toulon, 1912

In 1960, after a severe cyclone, the Greek ship *M D Alpine* was stranded on the shores of Sitakunda, Chittagong. It could not be re-floated and so remained there for several years. In 1965, the then in East Pakistan, Chittagong Steel House bought the ship and had it scrapped. It took years to scrap the vessel, but the work gave birth to the industry in Bangladesh. Until 1980 Gadani ship-breaking yard of Pakistan was the largest ship-breaking yard of the world.

Tightening environmental regulations resulted in increased costs of hazardous waste disposal in industrialised countries in the 1980s, causing ships to be exported to lower income nations, chiefly South Asia. This, in turn, created a far worse environmental problem, subsequently leading to the Basel Convention. In 2004 a Basel Convention decision officially classified old ships as "toxic waste", preventing them from leaving a country without the permission of the importing state. This has led to a resurgence of recycling in environmentally-compliant locations in developed countries, especially in former ship building yards.

On 31 December 2005, the French Navy's *Clemenceau* left Toulon to be dismantled in Alang, India despite protests over improper disposal capabilities and facilities for the toxic wastes. On 6 January 2006 the Supreme Court of India temporarily denied access to Alang, and the *Conseil d'État* ordered *Clemenceau* to return to French waters. Able UK in Hartlepool received a new disassembly contract to use accepted practices in scrapping the ship. The dismantling started on 18 November 2009 and the break-up was completed by the end of 2010, and the event was considered a turning point in the treatment of redundant vessels. Europe and the United States have actually had a resurgence in ship scrapping since the 1990s.

In 2009 the Bangladesh Environmental Lawyers Association won a legal case prohibiting all substandard ship breaking. For 14 months the industry could not import ships and thousands of jobs were lost before the ban was annulled. That same year, the global recession and lower demand for goods led to an increase in the supply of ships for decommissioning. The rate of scrapping is inversely correlated to the freight price, which collapsed in 2009.

Technique

The decommissioning process is entirely different in developed and developing countries. Both start with an auction for which the highest bidder wins the contract. The ship-breaker then acquires the vessel from the international broker who deals in outdated ships. The price paid is around $400 per tonne ($4–10 million), and the poorer the environmental legislation the higher the price. The purchase of water-craft makes up 69% of the income earned by the industry in Bangladesh, versus 2% for labour costs. The boat is taken to the decommissioning location either under its own power or with the use of tugs.

An aerial view of Chittagong, Bangladesh

Developing Countries

In developing countries, chiefly the Indian subcontinent, ships are run ashore on gently sloping sand tidal beaches at high tide so that they can be accessed for disassembly. As described in *"History"* (above), the sizeable ship breaking industry of Bangladesh traces its origin to a ship beached there accidentally during a cyclone. Manoeuvring a large ship onto a beach at high speed takes skill and daring even for a specialist captain, and is not always successful. Next, the anchor is dropped to steady the ship and the engine is shut down. It takes 50 labourers about three months to break down a normal-sized cargo vessel of about 40,000 tonnes.

The decommissioning begins with the draining of fuel and fire fighting liquid, which is sold to the trade. Any re-usable items—wiring, furniture and machinery—are sent to local markets or the trade. Unwanted materials become inputs to their relevant waste streams. Often, in lower income nations, these industries are no better than ship break-ing. For example, the toxic insulation is usually burnt off copper wire to access the met-al. Some crude safety precautions exist—chickens are lowered into the chambers of the ship, and if the birds return alive, they are considered safe. Workers also do not have proper clothing, footwear and masks.

Gas cutting in Chittagong, Bangladesh

Sledgehammers and oxy-acetylene gas-torches are used to cut up the steel hull. Cranes are not typically used on the ship, because of costs. Pieces of the hull simply fall off and are dragged inland, possibly aided with a winch or bulldozer. These are then cut into smaller pieces away from the coast. 90% of the steel is re-rollable scrap: higher quality steel plates that are heated and reused as reinforcement bar for construction. The remainder is transported to electric arc furnaces to be melted down into ingots for re-rolling mills. In the re-rolling mills, the heating of painted steel plates (in particular, those painted with chlorinated rubber paints) generates dioxins. Substances which are costly to dispose of, such as hazardous waste, are left on the beach or set on fire, even old batteries and half-empty cans of paint. Stockpiled in Bangladesh, for example, are 79,000 tonnes of asbestos, 240,000 tonnes of PCBs and 210,000 tonnes of Ozone-de-pleting substances.

Developed Countries

In developed countries the dismantling process should mirror the technical guidelines for the environmentally sound management of the full and partial dismantling of ships, published by the Basel Convention in 2003. Recycling rates of 98% can be achieved in these facilities.

Prior to dismantling, an inventory of dangerous substances should be compiled. All hazardous materials and liquids, such as bilge water, are removed before ship breaking. Holes should be bored for ventilation and all flammable vapours are extracted.

Vessels are initially taken to a dry dock or a pier, although a dry dock is considered more environmentally friendly because all spillage is contained and can easily be cleaned up.

Floating is, however, cheaper than a dry dock. Storm water discharge facilities will stop an overflow of toxic liquid into the waterways. The carrier is then secured to ensure its stability. Often the propeller is removed beforehand to allow the water-craft to be moved into shallower water.

Workers must completely strip the ship down to a bare hull, with objects cut free using saws, grinders, abrasive cutting wheels, hand held shears, plasma and gas torches. Anything of value, such as spare parts and electronic equipment is sold for re-use, although labour costs mean that low value items are not economical to sell. The Basel Convention demands that all yards separate hazardous and non-hazardous waste and have appropriate storage units, and this must be done before the hull is cut up. Asbestos, found in the engine room, is isolated and stored in custom-made plastic wrapping prior to being placed in secure steel containers, which are then landfilled.

Many hazardous wastes can be recycled into new products. Examples include lead-acid batteries or electronic circuit boards. Another commonly used treatment is cement based solidification and stabilization. Cement kilns are used because they can treat a range of hazardous wastes by improving physical characteristics and decreasing the toxicity and transmission of contaminants. A hazardous waste may also be "destroyed" by incinerating it at a high temperature; flammable wastes can sometimes be burned as energy sources. Some hazardous waste types may be eliminated using pyrolysis in a high temperature electrical arc, in inert conditions to avoid combustion. This treatment method may be preferable to high temperature incineration in some circumstances such as in the destruction of concentrated organic waste types, including PCBs, pesticides and other persistent organic pollutants. Dangerous chemicals can also be permanently stored in landfills as long as leaching is prevented.

Valuable metals, such as copper in electric cable, that are mixed with other materials may be recovered by the use of shredders and separators in the same fashion as e-waste recycling. The shredders cut the electronics into metallic and non-metallic pieces. Metals are extracted using magnetic separators, air flotation separator columns, shaker tables or eddy currents. The plastic almost always contains regulated hazardous waste (e.g., asbestos, PCBs, hydrocarbons) and cannot be melted down.

Large objects, such as engine parts, are extracted and sold as they become accessible. The hull is cut into 300 tonne sections, starting with the upper deck and working slowly downwards. While oxy-acetylene gas-torches are most commonly used, detonation charges can quickly remove large sections of the hull. These sections are transported to an electric arc furnace to be melted down into new ferrous products, though toxic paint must be stripped prior to heating.

Health Risks

70% of ships are simply run ashore in developing countries for disassembly, where (particularly in older vessels) asbestos, lead, polychlorinated biphenyls and heavy metals pose a danger for the workers. Burns from explosions and fire; suffocation; mutilation from falling metal; cancer and disease from toxins are regular occurrences in the industry. Asbestos was used heavily in ship construction until it was finally banned in most of the developed world in the mid-1980s. Currently, the costs associated with removing asbestos, along with the potentially expensive insurance and health risks, have meant that ship breaking in most developed countries is no longer economically viable. Dangerous vapors and fumes from burning materials can be inhaled, and dusty asbestos-laden areas are commonplace.

Removing the metal for scrap can potentially cost more than the value of the scrap metal itself. In the developing world, however, shipyards can operate without the risk of personal injury lawsuits or workers' health claims, meaning many of these shipyards may operate with high health risks. Protective equipment is sometimes absent or inadequate. The sandy beaches cannot sufficiently support the heavy equipment, which is thus prone to collapse. Many are injured from explosions when flammable gas is not removed from fuel tanks. In Bangladesh, a local watchdog group claims that one worker dies a week and one is injured per a day on average.

The problem is caused by negligence from national governments, shipyard operators, and former ship owners disregarding the Basel Convention. According to the Institute for Global Labour and Human Rights, workers who attempt to unionize are fired and then blacklisted. The employees have no formal contract or any rights, and sleep in over-crowded hostels. The authorities produce no comprehensive injury statistics, so the problem is underestimated. Child labour is also widespread: 20% of Bangladesh's ship breaking workforce are below 15 years of age, mainly involving in cutting with gas torches. Greenpeace discovered that the yard owners and local authorities are complicit in attempting to cover up the problem in order to reap the economic benefits.

There is, however, an active ship breaker's union in Mumbai, India (Mumbai Port Trust Dock and General Employees' Union) since 2003 with 15,000 members, which strikes to ensure fatality compensation. It has set up a sister branch in Alang, gaining paid holidays and safety equipment for workers since 2005. They hope to expand all along the South Asian coastline.

Several United Nations committees are increasing their coverage of ship breakers' human rights. In 2006, the International Maritime Organisation developed legally binding global legislation which concerns vessel design, vessel recycling and the enforcement of regulation thereof and a 'Green Passport' scheme. Water-craft must have an inventory of hazardous material before they are scrapped, and the facilities must meet health & safety requirements. The International Labour Organization created a voluntary set of guidelines for occupational safety in 2003. Nevertheless, Greenpeace found that even pre-existing mandatory regulation has had little noticeable effect for labourers, due to government corruption, yard owner secrecy and a lack of interest from countries who prioritise economic growth. To safeguard worker health, the report recommends that developed countries create a fund to support their families, certify carriers as 'gas-free' (i.e. safe for cutting) and to remove toxic materials in appropriate facilities before export. To supplement the international treaties, organisations such as the NGO Shipbreaking Platform, the Institute for Global Labour and Human Rights and ToxicsWatch Alliance are lobbying for improvements in the industry. There are also guards who look out for any reporters.

Environmental Risks

In recent years, ship breaking has become an issue of environmental concern beyond the health of the yard workers. Many ship breaking yards operate in developing nations with lax or no environmental law, enabling large quantities of highly toxic materials to escape into the general environment and causing serious health problems among ship breakers, the local population, and wildlife. Environmental campaign groups such as Greenpeace have made the issue a high priority for their activities.

Along the Indian subcontinent, ecologically-important mangrove forests, a valuable source of protection from tropical storms and monsoons, have been cut down to provide space for water-craft disassembly. In Bangladesh, for example, 40,000 mangrove trees were illegally chopped down in 2009. The World Bank has found that the country's beaching locations are now at risk from sea level rise. 21 fish and crustacean species have been wiped out in the country as a result of the industry as well. Lead, organotins such as tributyltin in anti-fouling paints, polychlorinated organic compounds, by-products of combustion such as polycyclic aromatic hydrocarbons, dioxins and furans are found in ships and pose a great danger to the environment.

The Basel Convention on the Control of Trans-boundary Movements of Hazardous Wastes and Their Disposal of 1989 has been ratified by 166 countries, including India and Bangladesh, and in 2004, End of Life Ships were subjected to its regulations. It aims to stop the transportation of dangerous substances to less developed countries and mandate the use of regulated facilities. However, Greenpeace reports that neither vessel exporter nor breaking countries are adhering to its policies. The organisation recommends that all parties enforce the Basel Convention in full, and hold those who break it liable. Furthermore, the decision to scrap a ship is often made in international waters, where the convention has no jurisdiction.

The Hong Kong Convention is a compromise. It allows ships to be exported for recycling, as long as various stipulations are met: All water-craft must have an inventory and every shipyard needs to publish a recycling plan to protect the environment. The Hong Kong Convention was adopted in 2009 but with few countries signing the agreement.

In March 2012 the European Commission proposed tougher regulations to ensure all parties take responsibility. Under these rules, if a vessel has a European flag, it must be disposed of in a shipyard on an EU "green list." The facilities would have to show that they are compliant, and it would be regulated internationally in order to bypass corrupt local authorities. However, there is evidence of ship owners changing the flag to evade the regulations. China's scrap industry has vehemently protested against the proposed European regulations. Although Chinese recycling businesses are less damaging than their South Asian counterparts, European and American ship breakers comply with far more stringent legislation.

List of Ship Breaking Yards

The following are some of world's largest ship breaking yards:

Bangladesh

- Chittagong Ship Breaking yard

China

- Changjiang Ship Breaking yard, located in Jiangyin, China

India

- Alang-Sosiya Ship Breaking Yard

Pakistan

- Gadani Ship Breaking yard

Turkey

- Aliaga Ship Breaking Yard

United States

- Esco Marine, Brownsville, Texas
- International Shipbreaking, Brownsville, Texas

United Kingdom

- Able UK, Graythorpe Dock, Teeside

Belgium

- Galloo, Ghent, formerly Van Heyghen Recycling

Ferrous Metal Recycling

Ferrous metals are able to be recycled, with steel being one of the most recycled materials in the world,. Ferrous metals contain an appreciable percentage of iron and the addition of carbon and other substances creates steel.

A pile of steel scrap in Brussels, waiting to be recycled

Description

In the United States, steel containers, cans, automobiles, appliances, and construction materials contribute the greatest weight of recycled materials. For example, in 2008, more than 97% of structural steel and 106% of automobiles were recycled, comparing the current steel consumption for each industry with the amount of recycled steel being produced (the late 2000s recession and the associated sharp decline in automobile production in the US explains the over-100% calculation). A typical appliance is about 75% steel by weight and automobiles are about 65% steel and iron.

The Universal Symbol for Recyclable Steel

The steel industry has been actively recycling for more than 150 years, in large part because it is economically advantageous to do so. It is cheaper to recycle steel than to mine iron ore and manipulate it through the production process to form new steel. Steel does not lose any of its inherent physical properties during the recycling process, and has drastically reduced energy and material requirements compared with refinement from iron ore. The energy saved by recycling reduces the annual energy consumption of the industry by about 75%, which is enough to power eighteen million homes for one year. According to the International Resource Panel's Metal Stocks in Society report, the per capita stock of steel in use in Australia, Canada, the European Union EU15, Norway, Switzerland, Japan, New Zealand and the US combined is 7,085 kilograms (15,620 lb) (about 860 million people in 2005).

The CEN Symbol for Recyclable Steel

Basic oxygen steelmaking (BOS) uses 25–35% recycled steel to make new steel. BOS steel usually contains lower concentrations of residual elements such as copper, nickel and molybdenum and is therefore more malleable than electric arc furnace (EAF) steel and is often used to make automotive fenders, tin cans, industrial drums or any product with a large degree of cold working. EAF steelmaking uses almost 100% recycled steel. This steel contains greater concentrations of residual elements that cannot be removed through the application of oxygen and lime. It is used to make structural beams, plates, reinforcing bar and other products that require little cold working. Downcycling of steel by hard-to-separate impurities such as copper or tin can only be prevented by well-aimed scrap selection or dilution by pure steel. Recycling one metric ton (1,000 kilograms) of steel saves 1.1 metric tons of iron ore, 630 kilograms of coal, and 55 kilograms of limestone.

Types of Scrap Used in Steelmaking

- Heavy melting steel – Industrial or commercial scrap steel greater than 6mm thick, such as plates, beams, columns, channels; may also include scrap machinery or implements or certain metal stampings

- Old car bodies – Vehicles with or without interiors and their original wheels

- Cast iron – Cast iron bathtubs, machinery, pipe and engine blocks

- Pressing steel – Domestic scrap metal up to approx. 6mm thick. Examples - White goods (fridges, washing machines, etc.), roofing iron, water heaters, water tanks and sheet metal offcuts

- Re-inforcing bars or mesh – Used in the construction industry within concrete structures

- Turnings – Remains of drilling or shaping steels. Also known as "borings" or "swarf"

- Manganese steel – Non magnetic, hardened steel used in the mining industry, cement mixers, rock crushers, and other high impact and abrasive environments.

- Rails – Rail or tram tracks

Recycling by Country

United States

As of 2008, more than 83% of steel was recycled in the United States. It is the most widely recycled material; in 2000, more than 60 million metric tons were recycled.

Precycling

Precycling is the practice of reducing waste by attempting to avoid bringing items which will generate waste into home or business. The U.S. Environmental Protection Agency (EPA) also cites that precycling is the preferred method of integrated solid waste management because it cuts waste at its source and therefore trash is eliminated before it is created. According to the EPA, precycling is also characterized as a decision-making process on the behalf of the consumer because it involves making informed judgments regarding a product's waste implications. The implications that are taken into consideration by the consumer include: whether a product is reusable, durable, or repairable; made from renewable or non-renewable resources; over-packaged; and whether or not the container is reusable.

About

Precycling has the ability to build industrial, social, environmental, and economic circumstances that allow for old products to be converted into new resources

- Industrial: increasing the independence from accumulative substances, such as heavy metals, fossil fuels, synthetics, etc.

- Economic: create a circular economy

- Ecological/environmental: allowance for more extensive and diverse natural habitats where the resources are returned to nature

- Societal: extend the capacity of precycling to meet everyone's needs

The concept of 'precycling' was coined in 1988 by social marketing executive Maureen O'Rorke in a public waste education campaign for the City of Berkeley. The application of precycling is not limited to large corporations, but can be administered on smaller scales in local communities. The reason precycling is effective on large scales and on small scales stems from the idea that it shares a common language between experts and non-experts, buyers and sellers, economists and environmentalists. However, it is important to consider that waste prevention systems, such as precycling, require the collaborative effort from several working parts. These parts include prevention targets, producer responsibility, householder charging, funding for pilot projects, public involvement, engagement of private and third sectors, and public campaigns that spread awareness.

Integration of Waste Management

The original three-pronged push for trash management is "Reduce, Reuse, Recycle." Precycling emphasizes "reducing and reusing", while harnessing and questioning the momentum and popularity of the term "recycle." In addition to this three-pronged waste management strategy of "Reduce, Reuse, Recycle", precycling incorporates four supplementary R's: Repair, Recondition, Remanufacture and Refuse. Waste is a resource that can be reused, recycled, recovered, or treated. Precycling differs from other singular forms of waste prevention because it encompasses not one behavior, but many.

Reduce

Reduce is a form of precycling that allows for the preservation of natural resources and also saves money on behalf of the manufacturer, the consumer, and the waste manager. Moreover, effective source reduction slows the depletion of environmental resources, prolongs the life of waste management facilities, and makes combustion and landfills safer by removing toxic waste components.

Reuse

Reuse is a form of precycling that reinvents items after their initial life and avoids creating additional waste.

Recycle

Although precycling harnesses the familiarity of the term recycling, it is still important to note the difference between recycling and prevention. Since precycling focuses on the prevention of waste production, this entails that measures are taken before a substance, material, or product has become waste. Whereas recycling is a type of precycling that involves taking action before existing waste is abandoned to nature. Recycling is a process where discarded materials are collected, sorted, processed, and used in the production of new products. Every time a person engages in the act of recycling, they help increase the market and bring the cost down. However, current research from the American Plastics Council states that only 25% of the nation's recycling capabilities are being utilized.

Traditionally recycling requires large amounts of energy to "melt down" and then re-manufacture items. While this may cut down on the amount of trash that is going into landfills, it is not sustainable unless the underlying energy supply is sustainable. In addition, recycling often means downcycling and always involves at least some loss of the original material, so primary extraction is still required to make up the difference. Precycling reduces these problems by using less material in the first place, so less has to be recycled.

Repairing

Repair is a type of precycling that corrects specified faults in a product, however the quality of a repaired product is inferior to reconditioned or remanufactured items. One survey found that 68% of the respondents believed repairing was not cost efficient and sought alternative methods such as reconditioning or remanufacturing.

Reconditioning

Reconditioning is a type of precycling that requires the rebuilding of major components to restore a product's working condition, which is expected to be inferior to the original product.

Remanufacturing

Remanufacturing is another type of precycling that involves the greatest degree of work content, which results in superior quality of the product. In order to remanufacture a product, it requires a total dismantling of the product and the subsequent restoration and replacement of its parts. Remanufacturing is a preferred method of waste reduction compared to repairing and reconditioning because it preserves the embodied energy that has been used to shape the components of a product for their first life and it only requires 20-25% of the initial energy used in formation.

Refuse

Refusal to buy certain products due to detrimental impacts on the environment or wasteful packaging is another type of precycling because the rejection of such items paves the way for products that can be reduced, reused, or recycled.

Zero-waste Strategy

A zero waste approach aims to prevent rather than just reduce accumulated waste. Zero-waste goes beyond recycling to include the whole system, which includes the flow of resources and waste through human society. This "design principle" works to maximize recycling, minimize waste, reduce consumption and ensures that products are reused, repaired or recycled back into nature or the market. This preventative approach is more manageable and effective than incremental approaches that focus on gradually reducing the amount of impact because it is less complex and contains less information, which permits wider public participation.

Sustainability

In regards to sustainability, the term itself is often associated with resource constraints and maintenance of the status quo rather than growth and prosperity. However, with the implementation of a zero-waste management strategy, sustainable practices can push the status quo in order to create a society that is capable of development, technically and culturally advanced, dynamic in population and production, thoughtful with the use of non-renewable resources, and diverse, democratic, and challenging.

Economic Effects

Increased waste production is often negatively associated with increased economic growth. However, a zero-waste management strategy allows for economic growth that works cohesively with sustainability rather than against it. The implementation of a zero-waste strategy is part of an economic goal-set that aims to create a circular economy. A circular economy refers to a closed-loop socio-economic system that focuses on minimizing wastes while simultaneously maximizing stocks of resources for the economy. This closed-loop design diverts linear (open-loop) waste disposal streams into new raw material streams.

In a circular economy, one way to minimize waste is through the employment of precycling insurance, which allows for a full range of financed waste prevention opportunities. This type of insurance would set premiums related to the risk of a product ending up as waste, and these premiums would serve to fund actions concerning waste prevention. When establishing a premium for precycling insurance several factors need to calculated: recyclability or biodegradability; provision of infrastructure, habitat or collaborations for the generation of the product from new resources; the ecosystem concentrations of product components above natural levels. The idea of precycling insurance is plausible considering the aim of insurance industries is to avoid losses rather than paying for losses. However, in order for this idea to work, private and third sectors need to be involved and engaged in the issue. In this instance, a third sector refers to small charities and a handful of societal enterprises that coordinate with charity shops.

Environmental Effects

According to the "Extended Producer Responsibility" principle, impacts are substantially determined at the point of design where key decisions are made on materials, production process, and how products are used and disposed of at the end of life-cycle, which falls on the producer. However, in a circular economy there is the recognition that nature's capacity needs to be maximized through the reprocess of biodegradable wastes produced by industries and human activity. This task is accomplished through the procurement and funding of precycling insurance premiums that invest in systematic preservation of endangered habitats, careful harvesting of biological resources and expansions of productive ecosystems. Additionally, in terms of climate change, precycling insurance offers a flexible alternative to the binding limits on greenhouse gas emissions and international taxation on mineral fuels. In terms of waste management systems, the environment benefits from the reparation of products to the greatest degree because less energy is required and the majority of the original material is kept intact.

Societal Effects

The social structure operating under a circular economy is referred to as a circular society. The aim of a circular society is to create a cooperative culture by means of problem-prevention, resource-availability and fuller participation, with reference to precycling. One critique of this approach, in terms of waste management, is that It is difficult to maintain a cooperative culture within a society because they it is constantly evolving and changing.

Raising Awareness

There is an increasing public awareness on the need for sustainable production and consumption. One campaign that aimed at raising awareness of precycling focused on whether people's self-reported behaviors were affected by exposure to precycling advertisements on the radio, television, or in-store flyers. The researchers concluded that the most effective results stemmed from the inclusion of social rewards that invoke an intrinsic motivation to engage in precycling behaviors.

Another way to raise awareness is through statistics that highlight the potential impacts that can be achieved through waste prevention. For instance, if 70 million Americans bought a half-gallon container of milk each week (instead of two quarts), then 41.6 million pounds of paper discards and 5.7 pounds of plastic discards would be reduced annually. This transition from two quarts to a half-gallon would save $145.6 million on packaging each year.

Implementation

In order to effectively implement precycling practices and behaviors, the public needs to feel "enabled", "engaged", "encouraged", and "exemplified" in their efforts to partake in precycling.

Not only can the average consumer practice precycling, but industries can also participate. Purchasing from parts suppliers, reuse of chemicals, and reduction of unnecessary packaging are some methods. There are some companies and countries that have taken it upon themselves to implement more sustainable practices that align with precycling principles. For instance, Fonterra reduced its packaging through the implementation of bulking, reuse and redesign. Further, Waste Management New Zealand created Recycle New Zealand, which provided a subsidiary focusing on the collection of materials that could be diverted and sorted prior to the operations of reducing, recycling, or recovering. Moving forward, free-trade organizations can further implement precycling practices by exploring this strategy as a new way to reduce regulations and to promote greater industrial freedom of choice.

Moreover, the individual consumer can develop precycling habits by engaging in the following practices and behaviors:

"Enviro-shopping"

Enviro-shopping is considered shopping with the environment and implements a precycling strategy:

- Bringing one's own grocery bag or bring old ones back to the store

- Buying packages with the least amount of packaging

- Buying in bulk, but not buying more than one will use

- Looking for products with reusable dishes

Product Selection

Products to choose from in accordance with precycling principles:

- Plastic milk jugs or glass milk containers (no cartons)

- Fresh fruit and vegetables

- Concentrated products that involves less packaging

- Recycled products

- Rechargeable batteries

Behaviors

In addition to shopping practices that implement precycling principles, there are also behaviors that can be undertaken to prevent waste:

- Home composting

- Avoid junk mail

- Buy second-hand

- One way to participate in precycling is to carry a "precycling kit". Include a Tupperware or non-disposable container, silverware set, a cloth napkin or handkerchief, and a thermos or water-bottle within a cloth bag that can double as a grocery/shopping bag.

References

- The League of Women Voters (1993). The Garbage Primer. New York: Lyons & Burford. pp. 35–72. ISBN 1-55821-250-7.

- Cleveland, Cutler J.; Morris, Christopher G. (November 15, 2013). Handbook of Energy: Chronologies, Top Ten Lists, and Word Clouds. Elsevier. p. 461. ISBN 978-0-12-417019-3.

- Dadd-Redalia, Debra (January 1, 1994). Sustaining the earth: choosing consumer products that are safe for you, your family, and the earth. New York: Hearst Books. p. 103. ISBN 978-0-688-12335-2.

- Carl A. Zimring (2005). Cash for Your Trash: Scrap Recycling in America. New Brunswick, NJ: Rutgers University Press. ISBN 0-8135-4694-X.

- Lynn R. Kahle; Eda Gurel-Atay, eds. (2014). Communicating Sustainability for the Green Economy. New York: M.E. Sharpe. ISBN 978-0-7656-3680-5.

- Huesemann, M.; Huesemann, J. (2011). Techno-fix: Why Technology Won't Save Us or the Environment. New Society Publishers. p. 464. ISBN 978-0-86571-704-6. Retrieved 2016-07-07.

- The Self-Sufficiency Handbook: A Complete Guide to Greener Living by Alan Bridgewater pg. 62--Skyhorse Publishing Inc., 2007 ISBN 1-60239-163-7, ISBN 978-1-60239-163-5

- Sahni, S.; Gutowski, T. G. (2011). "Your scrap, my scrap! The flow of scrap materials through international trade" (PDF). IEEE International Symposium on Sustainable Systems and Technology (ISSST): 1–6. doi:10.1109/ISSST.2011.5936853. ISBN 978-1-61284-394-0.

- Nguemaleu, Raoul-Abelin Choumin; Montheu, Lionel (2014-05-09). Roadmap to Greener Computing. CRC Press. p. 170. ISBN 9781466506848.

- "Report: "On the Making of Silk Purses from Sows' Ears," 1921: Exhibits: Institute Archives & Special Collections: MIT". mit.edu. Retrieved July 7, 2016.

- Lohr, Steve (1993-04-14). "Recycling Answer Sought for Computer Junk". The New York Times. ISSN 0362-4331. Retrieved 2015-07-29.

- "Bulgaria opens largest WEEE recycling factory in Eastern Europe". www.ask-eu.com. WtERT Germany GmbH. 12 Jul 2010. Retrieved 2015-07-29.

- "EnvironCom opens largest WEEE recycling facility / waste & recycling news". www.greenwise-business.co.uk. The Sixty Mile Publishing Company. 4 March 2010. Retrieved 2015-07-29.

- Goodman, Peter S. (11 Jan 2012). "Where Gadgets Go To Die: E-Waste Recycler Opens New Plant In Las Vegas". The Huffington Post. Retrieved 2015-07-29.

- Moses, Asher (19 Nov 2008). "New plant tackles our electronic leftovers - BizTech - Technology - smh.com.au". www.smh.com.au. Retrieved 2015-07-29.

Important Measures of Waste Management

The withdrawal of disposed materials for recycling to obtain full benefits is termed as resource recovery and the organizations that promote eco-innovation is industrial symbiosis. Other measures of waste management are eco-industrial park, pyrolysis, waste hierarchy and resource efficiency. The text strategically encompasses and incorporates the major components and key concepts of waste management.

Resource Recovery

Resource recovery is the selective extraction of disposed materials for a specific next use, such as recycling, composting or energy generation in order to extract the maximum benefits from products, delay the consumption of virgin resources, and reduce the amount of waste generated. Resource recovery differs from the management of waste by using life-cycle analysis (LCA) to offer alternatives to landfill disposal of discarded materials. A number of studies on municipal solid waste (MSW) have indicated that administration, source separation and collection followed by reuse and recycling of the non-organic fraction and energy and compost/fertilizer production of the organic waste fraction via anaerobic digestion to be the favoured alternatives to landfill disposal.

Similarly, in the context of sanitation, the term "resource recovery" is used to denote sanitation systems that aim to recover and reuse the resources that are contained in wastewater and excreta (urine and feces). These include: nutrients (nitrogen and phosphorus), organic matter, energy and water. This concept is also referred to as ecological sanitation or productive sanitation.

Recycling of Solid Waste

Recycling is a resource recovery practice that refers to the collection and reuse of disposed materials such as empty beverage containers. The materials from which the items are made can be reprocessed into new products. Material for recycling may be collected separately from general waste using dedicated bins and collection vehicles, or sorted directly from mixed waste streams.

The most common consumer products recycled include aluminium such as beverage cans, copper such as wire, steel food and aerosol cans, old steel furnishings or equipment, polyethylene and PET bottles, glass bottles and jars, paperboard cartons, news-papers, magazines and light paper, and corrugated fiberboard boxes.

Steel crushed and baled for recycling

PVC, LDPE, PP, and PS are also recyclable. These items are usually composed of a single type of material, making them relatively easy to recycle into new products. The recycling of complex products (such as computers and electronic equipment) is more difficult, due to the additional dismantling and separation required.

The type of recycling material accepted varies by city and country. Each city and country have different recycling programs in place that can handle the various types of recyclable materials.

Organic Matter

Disposed materials that are organic in nature, such as plant material, food scraps, and paper products, can be recycled using biological composting and digestion processes to decompose the organic matter. The resulting organic material is then recycled as mulch or compost for agricultural or landscaping purposes. In addition, waste gas from the process (such as methane) can be captured and used for generating electricity and heat (CHP/cogeneration) maximising efficiencies. The intention of biological processing is to control and accelerate the natural process of decomposition of organic matter.

An active compost heap.

There is a large variety of composting and digestion methods and technologies varying in complexity from simple home compost heaps, to small town scale batch digesters, industrial-scale enclosed-vessel digestion of mixed domestic waste. Methods of biological decomposition are differentiated as being aerobic or anaerobic methods, though hybrids of the two methods also exist.

Anaerobic digestion of the organic fraction of municipal solid waste (MSW) has been found to be in a number of LCA analysis studies to be more environmentally effective, than landfill, incineration or pyrolysis. The resulting biogas (methane) though must be used for cogeneration (electricity and heat preferably on or close to the site of production) and can be used with a little upgrading in gas combustion engines or turbines. With further upgrading to synthetic natural gas it can be injected into the natural gas network or further refined to hydrogen for use in stationary cogeneration fuel cells. Its use in fuel cells eliminates the pollution from products of combustion. There is a large variety of composting and digestion methods and technologies varying in complexity from simple home compost heaps, to small town scale batch digesters, industrial-scale, enclosed-vessel digestion of mixed domestic waste. Methods of biological decomposition are differentiated as being aerobic or anaerobic methods, though hybrids of the two methods also exist.

Recovery Methods

In many countries, source-separated curbside collection is one method of resource recovery.

Australia

In Australia, every urban domestic household is provided with three bins: one for recycling; another for general waste; and another for garden materials, this bin is provided by the municipality if requested. To encourage recycling, municipalities provide large recycle bins, which are larger than general waste bins. Many American localities have dual-stream recycling, with paper collected in bags or boxes and all other materials in a recycling bin. In either case, the recovered materials are trucked to a materials recovery facility for further processing.

Municipal, commercial and industrial, construction and demolition debris is dumped at landfills and some is recycled. Household disposal materials are segregated: recyclables sorted and made into new products, and unusable material is dumped in landfill areas. According to the Australian Bureau of Statistics (ABS), the recycling rate is high and is "increasing, with 99% of households reporting that they had recycled or reused within the past year (2003 survey), up from 85% in 1992". In 2002–03 "30% of materials from municipalities, 45% from commercial and industrial generators and 57% from construction and demolition debris" was recycled. Energy is produced is part of resource recovery as well:

some landfill gas is captured for fuel or electricity generation, although this is considered the last resort, as the point of resource recovery is avoidance of landfill disposal altogether.

Sustainability

Resource recovery is a key component in a business' ability to maintaining ISO14001 accreditation. Companies are encouraged to improve their environmental efficiencies each year. One way to do this is by changing a company from a system of managing wastes to a resource recovery system (such as recycling: glass, food waste, paper and cardboard, plastic bottles etc.)

Education and awareness in the area of resource recovery is increasingly important from a global perspective of resource management. The Talloires Declaration is a declaration for sustainability concerned about the unprecedented scale and speed of environmental pollution and degradation, and the depletion of natural resources. Local, regional, and global air pollution; accumulation and distribution of toxic wastes; destruction and depletion of forests, soil, and water; depletion of the ozone layer and emission of "green house" gases threaten the survival of humans and thousands of other living species, the integrity of the earth and its biodiversity, the security of nations, and the heritage of future generations. Several universities have implemented the Talloires Declaration by establishing environmental management and resource recovery programs. University and vocational education are promoted by various organizations, e.g., WAMITAB and Chartered Institution of Wastes Management. Many supermarkets encourage customers to use their reverse vending machines to deposit used purchased containers and receive a refund from the recycling fees. Brands that manufacture such machines include Tomra and Envipco.

In 2010, CNBC aired the documentary *Trash Inc: The Secret Life of Garbage* about waste, what happens to it when it's "thrown away", and its impact on the world.

Extended Producer Responsibility

Extended producer responsibility (EPR) is a strategy designed to promote the integration of all costs associated with products throughout their life cycle (including end-of-life disposal costs) into the market price of the product. Extended producer responsibility is meant to impose accountability over the entire lifecycle of products and packaging introduced to the market. This means that firms which manufacture, import and/or sell products are required to be responsible for the products after their useful life as well as during manufacture.

Industrial Symbiosis

Industrial Symbiosis a subset of industrial ecology. It describes how a network of diverse organizations can foster eco-innovation and long-term culture change, create and share mutually profitable transactions - and improve business and technical processes.

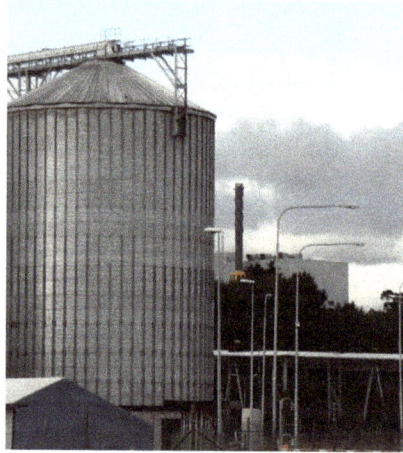

Example of Industrial symbiosis: waste steam from a waste incinerator (right) is piped to an ethanol plant (left) where it is used as an input to their production process

Although geographic proximity is often associated with industrial symbiosis, it is neither necessary nor sufficient—nor is a singular focus on physical resource exchange. In practice, using industrial symbiosis as an approach to commercial operations – using, recovering and redirecting resources for reuse – results in resources remaining in productive use in the economy for longer. This in turn creates business opportunities, reduces demands on the earth's resources, and provides a stepping-stone towards creating a circular economy. The industrial symbiosis model devised and managed by International Synergies Limited is a facilitated model operating at the national scale in the United Kingdom (NISP - National Industrial Symbiosis Programme), and at other scales around the world. International Synergies Limited has developed global expertise in IS, instigating programmes in Belgium, Brazil, Canada, China, Denmark, Finland, Hungary, Italy, Mexico, Poland, Romania, Slovakia, South Africa and Turkey, as well as the UK. Industrial symbiosis is a subset of industrial ecology, with a particular focus on material and energy exchange. Industrial ecology is a relatively new field that is based on a natural paradigm, claiming that an industrial ecosystem may behave in a similar way to the natural ecosystem wherein everything gets recycled.

Introduction

Eco-industrial development is one of the ways in which industrial ecology contributes to the integration of economic growth and environmental protection. Some of the examples of eco-industrial development are:

- Circular economy (single material and/or energy exchange)

- Greenfield eco-industrial development (geographically confined space)

- Brownfield eco-industrial development (geographically confined space)

- Eco-industrial network (no strict requirement of geographical proximity)

- Virtual eco-industrial network (networks spread in large areas e.g. regional network)

- Networked Eco-industrial System (macro level developments with links across regions)

"This classification omits any industrial sector-wide approaches and appreciates the diversity of the industrial system which is a key feature of industrial symbiosis. It is aimed to include initiatives that focus on achieving utility sharing and symbiosis among diverse sectors of industry". It is the diversity and the openness of industrial symbiosis that makes it a unique approach to eco-industrial development.

Industrial symbiosis engages traditionally separate industries in a collective approach to competitive advantage involving physical exchange of materials, energy, water, and/ or by-products. The keys to industrial symbiosis are collaboration and the synergistic possibilities offered by geographic proximity". The sharing of information is even more critical with the emergence of virtual globes such as Google Earth. These tools can greatly simplify the geographical analysis involved in determining potential IS opportunities.

Industrial symbiosis systems collectively optimize material and energy use at efficiencies beyond those achievable by any individual process alone. IS systems such as the web of materials and energy exchanges among companies in Kalundborg, Denmark have spontaneously evolved from a series of micro innovations over a long time scale; however, the engineered design and implementation of such systems from a macro planner's perspective, on a relatively short time scale, proves challenging. Nevertheless, there are examples of industrial symbiosis being approached as national / regional initiatives with some significant success particularly in Europe.

Often, access to information on available by-products is non-existent. These by-products are considered waste and typically not traded or listed on any type of exchange.

Example

Recent work reviewed government policies necessary to construct a multi-gigaWatt photovoltaic factory and complementary policies to protect existing solar companies are outlined and the technical requirements for a symbiotic industrial system are explored to increase the manufacturing efficiency while improving the environmental impact of solar photovoltaic cells. The results of the analysis show that an eight-factory industrial symbiotic system can be viewed as a medium-term investment by any government, which will not only obtain direct financial return, but also an improved global environment. This is because synergies have been identified for co-locating glass manufacturing and photovoltaic manufacturing. The waste heat from glass manufacturing can be used in industrial-sized greenhouses for food production. Even within the PV plant itself a secondary chemical recycling plant can reduce environmental impact while improving economic performance for the group of manufacturing facilities.

In DCM Shriram consolidated limited (Kota unit) produces Caustic Soda, calcium Carbide, Cement and PVC Resins.Chlorine and Hydrogen are obtained as by-products from caustic soda production, while Calcium carbide produced is partly sold and partly is treated with water to form Slurry(Aqueous solution of Calcium Hydroxide) and Ethylene. The chlorine and ethylene produced are utilised to form PVC compounds, while the slurry is consumed for Cement production by wet process. Hydrochloric Acid is prepared by direct synthesis where The pure chlorine gas can be combined with hydrogen to produce hydrogen chloride in the presence of UV light.

Eco-industrial Park

An eco-industrial park (EIP) is an industrial park in which businesses cooperate with each other and with the local community in an attempt to reduce waste and pollution, efficiently share resources (such as information, materials, water, energy, infrastructure, and natural resources), and help achieve sustainable development, with the intention of increasing economic gains and improving environmental quality. An EIP may also be planned, designed, and built in such a way that it makes it easier for businesses to co-operate, and that results in a more financially sound, environmentally friendly project for the developer.

View of the Kalundborg Eco-industrial Park

The Eco-industrial Park Handbook states that "An Eco-Industrial Park is a community of manufacturing and service businesses located together on a common property. Members seek enhanced environmental, economic, and social performance through collaboration in managing environmental and resource issues."

Based on the concepts of industrial ecology, collaborative strategies not only include by-product synergy ("waste-to-feed" exchanges), but can also take the form of waste-water cascading, shared logistics and shipping & receiving facilities, shared parking, green technology purchasing blocks, multi-partner green building retrofit, district energy systems, and local education & resource centres. This is an application of a systems approach, in which designs and processes/activities are integrated to address multiple objectives.

EIPs can be developed as greenfield land projects, where the eco-industrial intent is present throughout the planning, design and site construction phases, or developed through retrofits and new strategies in existing industrial developments.

Examples

"Industrial symbiosis" is a related but more limited concept in which companies in a region collaborate to utilize each other's by-products and otherwise share resources. In Kalundborg, Denmark a symbiosis network links a 1500MW coal-fired power plant with the community and other companies. Surplus heat from this power plant is used to heat 3500 local homes in addition to a nearby fish farm, whose sludge is then sold as a fertilizer. Steam from the power plant is sold to Novo Nordisk, a pharmaceutical and enzyme manufacturer, in addition to a Statoil plant. This reuse of heat reduces the amount thermal pollution discharged to a nearby fjord. Additionally, a by-product from the power plant's sulfur dioxide scrubber contains gypsum, which is sold to a wallboard manufacturer. Almost all of the manufacturer's gypsum needs are met this way, which reduces the amount of open-pit mining needed. Furthermore, fly ash and clinker from the power plant is utilized for road building and cement production.

The industrial symbiosis at Kalundborg was not created as a top-down initiative, but instead evolved gradually. As environmental regulations became stricter, firms were motivated reduce the cost of compliance, and turn their by-products into economic products.

In Canada, eco-industrial parks exist across the country and have enjoyed some success. The best known example is Burnside Park, in Halifax, Nova Scotia. With support from Dalhousie University's Eco-Efficiency Centre, the more than 1,500 businesses have been improving their environmental performance and developing profitable partnerships. Subsequently, two greenfield industrial developments have been started in Alberta: TaigaNova Eco-Industrial Park is in the heart of the Athabasca oil sands, while Innovista Eco-Industrial Park is a gateway to the Rocky Mountains ~300km west of Edmonton.

Other Usage

EIPs also refer to industrial parks where a "green" approach has been taken towards the infrastructure and development of the site. This can include green infrastructure related to Renewable Energy Systems; stormwater, groundwater and wastewater management; road surfaces; and transportation demand management. Green building practices can also be encouraged or mandated

EIPs are often used as a stimulus for economic diversification in the community or region where they are located. Anchor tenants, such as bio-based product manufacturers or waste-to-energy facilities, etc., can attract complementary businesses as suppliers, scavengers/recyclers, service providers, downstream users and other businesses that could benefit from eco-industrial strategies.

Suggested Usage

It is suggested that EIPs be used as a means of growing the renewable energy sector. In the case of a Solar Photovoltaic (PV) Manufacturing plant, an EIP can increase the manufacturing efficiency to make it more economical, while reducing the environmental impact of producing the solar cells. In essence, this assists the growth of the renewable energy industry and the environmental benefits that come with replacing fossil-fuels.

Pyrolysis

Pyrolysis is a thermochemical decomposition of organic material at elevated temperatures in the absence of oxygen (or any halogen). It involves the simultaneous change of chemical composition and physical phase, and is irreversible. The word is coined from the Greek-derived elements *pyro* "fire" and *lysis* "separating".

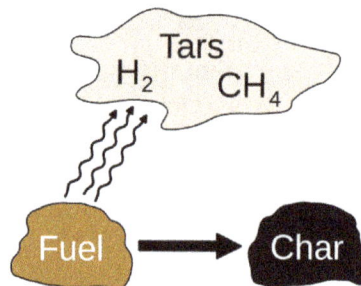

Simplified depiction of pyrolysis chemistry.

Pyrolysis is a type of thermolysis, and is most commonly observed in organic materials exposed to high temperatures. It is one of the processes involved in charring wood, starting at 200–300 °C (390–570 °F). It also occurs in fires where solid fuels are burning or when vegetation comes into contact with lava in volcanic eruptions. In general, pyrolysis of organic substances produces gas and liquid products and leaves a solid residue richer in carbon content, char. Extreme pyrolysis, which leaves mostly carbon as the residue, is called carbonization.

The process is used heavily in the chemical industry, for example, to produce charcoal, activated carbon, methanol, and other chemicals from wood, to convert ethylene dichloride into vinyl chloride to make PVC, to produce coke from coal, to convert biomass into syngas and biochar, to turn waste plastics back into usable oil, or waste into safely disposable substances, and for transforming medium-weight hydrocarbons from oil into lighter ones like gasoline. These specialized uses of pyrolysis may be called various names, such as dry distillation, destructive distillation, or cracking. Pyrolysis is also used in the creation of nanoparticles, zirconia and oxides utilizing an ultrasonic nozzle in a process called ultrasonic spray pyrolysis (USP).

Pyrolysis also plays an important role in several cooking procedures, such as baking, frying, grilling, and caramelizing. It is a tool of chemical analysis, for example, in mass spectrometry and in carbon-14 dating. Indeed, many important chemical substances, such as phosphorus and sulfuric acid, were first obtained by this process. Pyrolysis has been assumed to take place during catagenesis, the conversion of buried organic matter to fossil fuels. It is also the basis of pyrography. In their embalming process, the ancient Egyptians used a mixture of substances, including methanol, which they obtained from the pyrolysis of wood.

Pyrolysis differs from other processes like combustion and hydrolysis in that it usually does not involve reactions with oxygen, water, or any other reagents. In practice, it is not possible to achieve a completely oxygen-free atmosphere. Because some oxygen is present in any pyrolysis system, a small amount of oxidation occurs.

The term has also been applied to the decomposition of organic material in the presence of superheated water or steam (hydrous pyrolysis), for example, in the steam cracking of oil.

Occurrence and Uses

Fire

Pyrolysis is usually the first chemical reaction that occurs in the burning of many solid organic fuels, like wood, cloth, and paper, and also of some kinds of plastic. In a wood fire, the visible flames are not due to combustion of the wood itself, but rather of the gases released by its pyrolysis, whereas the flame-less burning of a solid, called smouldering, is the combustion of the solid residue (char or charcoal) left behind by pyrolysis. Thus, the pyrolysis of common materials like wood, plastic, and clothing is extremely important for fire safety and firefighting. In pyrolysis there is a gas phase present. It should not be confused with hydrothermal reactions such as hydrothermal gasification, hydrothermal liquidation, and hydrothermal carbonization, which occur in aqueous environments because the temperatures and reaction pathways differ, with ionic reactions favored in aqueous reactions and radical reactions favored in the absence of water.

Cooking

Pyrolysis occurs whenever food is exposed to high enough temperatures in a dry environment, such as roasting, baking, toasting, or grilling. It is the chemical process responsible for the formation of the golden-brown crust in foods prepared by those methods.

In normal cooking, the main food components that undergo pyrolysis are carbohydrates (including sugars, starch, and fibre) and proteins. Pyrolysis of fats requires a much higher temperature, and, since it produces toxic and flammable products (such as acrolein), it is, in general, avoided in normal cooking. It may occur, however, when grilling fatty meats over hot coals.

Even though cooking is normally carried out in air, the temperatures and environmental conditions are such that there is little or no combustion of the original substances or their decomposition products. In particular, the pyrolysis of proteins and carbohydrates begins at temperatures much lower than the ignition temperature of the solid residue, and the volatile subproducts are too diluted in air to ignite. (In flambé dishes, the flame is due mostly to combustion of the alcohol, while the crust is formed by pyrolysis as in baking.)

Pyrolysis of carbohydrates and proteins requires temperatures substantially higher than 100 °C (212 °F), so pyrolysis does not occur as long as free water is present, e.g., in boiling food — not even in a pressure cooker. When heated in the presence of water, carbohydrates and proteins suffer gradual hydrolysis rather than pyrolysis. Indeed, for most foods, pyrolysis is usually confined to the outer layers of food, and begins only after those layers have dried out.

Food pyrolysis temperatures are, however, lower than the boiling point of lipids, so pyrolysis occurs when frying in vegetable oil or suet, or basting meat in its own fat.

Pyrolysis also plays an essential role in the production of barley tea, coffee, and roasted nuts such as peanuts and almonds. As these consist mostly of dry materials, the process of pyrolysis is not limited to the outermost layers but extends throughout the materials. In all these cases, pyrolysis creates or releases many of the substances that contribute to the flavor, color, and biological properties of the final product. It may also destroy some substances that are toxic, unpleasant in taste, or those that may contribute to spoilage.

Controlled pyrolysis of sugars starting at 170 °C (338 °F) produces caramel, a beige to brown water-soluble product widely used in confectionery and (in the form of caramel coloring) as a coloring agent for soft drinks and other industrialized food products.

Solid residue from the pyrolysis of spilled and splattered food creates the brown-black encrustation often seen on cooking vessels, stove tops, and the interior surfaces of ovens.

Charcoal

Pyrolysis has been used since ancient times for turning wood into charcoal on an industrial scale. Besides wood, the process can also use sawdust and other wood waste products.

Charcoal is obtained by heating wood until its complete pyrolysis (carbonization) occurs, leaving only carbon and inorganic ash. In many parts of the world, charcoal is still produced semi-industrially, by burning a pile of wood that has been mostly covered with mud or bricks. The heat generated by burning part of the wood and the volatile by-products pyrolyzes the rest of the pile. The limited supply of oxygen prevents the charcoal from burning. A more modern alternative is to heat the wood in an airtight metal vessel, which is much less polluting and allows the volatile products to be condensed.

The original vascular structure of the wood and the pores created by escaping gases combine to produce a light and porous material. By starting with a dense wood-like material, such as nutshells or peach stones, one obtains a form of charcoal with particularly fine pores (and hence a much larger pore surface area), called activated carbon, which is used as an adsorbent for a wide range of chemical substances.

Biochar

Residues of incomplete organic pyrolysis, e.g., from cooking fires, are thought to be the key component of the terra preta soils associated with ancient indigenous communities of the Amazon basin. Terra preta is much sought by local farmers for its superior fertility compared to the natural red soil of the region. Efforts are underway to recreate these soils through biochar, the solid residue of pyrolysis of various materials, mostly organic waste.

Biochar improves the soil texture and ecology, increasing its ability to retain fertilizers and release them slowly. It naturally contains many of the micronutrients needed by plants, such as selenium. It is also safer than other "natural" fertilizers such as animal manure, since it has been disinfected at high temperature. And, since it releases its nutrients at a slow rate, it greatly reduces the risk of water table contamination.

Biochar is also being considered for carbon sequestration, with the aim of mitigation of global warming. The solid, carbon-containing char produced can be sequestered in the ground, where it will remain for several hundred to a few thousand years.

Coke

Pyrolysis is used on a massive scale to turn coal into coke for metallurgy, especially steel-making. Coke can also be produced from the solid residue left from petroleum refining.

Those starting materials typically contain hydrogen, nitrogen, or oxygen atoms combined with carbon into molecules of medium to high molecular weight. The coke-making or "coking" process consists of heating the material in closed vessels to very high temperatures (up to 2,000 °C or 3,600 °F) so that those molecules are broken down into lighter volatile substances, which leave the vessel, and a porous but hard residue that is mostly carbon and inorganic ash. The amount of volatiles varies with the source material, but is typically 25–30% of it by weight.

Carbon Fiber

Carbon fibers are filaments of carbon that can be used to make very strong yarns and textiles. Carbon fiber items are often produced by spinning and weaving the desired item from fibers of a suitable polymer, and then pyrolyzing the material at a high temperature (from 1,500–3,000 °C or 2,730–5,430 °F).

The first carbon fibers were made from rayon, but polyacrylonitrile has become the most common starting material.

For their first workable electric lamps, Joseph Wilson Swan and Thomas Edison used carbon filaments made by pyrolysis of cotton yarns and bamboo splinters, respectively.

Pyrolytic Carbon

Pyrolysis is the reaction used to coat a preformed substrate with a layer of pyrolytic carbon. This is typically done in a fluidized bed reactor heated to 1,000–2,000 °C or 1,830–3,630 °F. Pyrolytic carbon coatings are used in many applications, including artificial heart valves.

Biofuel

Pyrolysis is the basis of several methods that are being developed for producing fuel from biomass, which may include either crops grown for the purpose or biological waste products from other industries. Crops studied as biomass feedstock for pyrolysis include native North American prairie grasses such as *switchgrass* and bred versions of other grasses such as *Miscantheus giganteus*. Crops and plant material wastes provide biomass feedstock on the basis of their lignocellulose portions.

Although synthetic diesel fuel cannot yet be produced directly by pyrolysis of organic materials, there is a way to produce similar liquid (bio-oil) that can be used as a fuel, after the removal of valuable bio-chemicals that can be used as food additives or pharmaceuticals. Higher efficiency is achieved by the so-called flash pyrolysis, in which finely divided feedstock is quickly heated to between 350 and 500 °C (660 and 930 °F) for less than 2 seconds.

Fuel bio-oil can also be produced by hydrous pyrolysis from many kinds of feedstock, including waste from pig and turkey farming, by a process called thermal depolymerization (which may, however, include other reactions besides pyrolysis).

Adhesives

Neanderthals used pyrolysis of birch bark to produce a pitch with which they secured flaked stones to spear shafts. Recently, researchers have developed a process to pyrolyze birch bark to produce an oil that can replace phenol in phenol formaldehyde resin (these resins are mostly used to manufacture plywood).

Pesticides

Pyrolysis can also be used to produce pesticides from biomass.

Plastic Waste Disposal

Anhydrous pyrolysis can also be used to produce liquid fuel similar to diesel from plastic waste, with a higher cetane value and lower sulphur content than traditional diesel. Using pyrolysis to extract fuel from end-of-life plastic is a second-best option after recycling, is environmentally preferable to landfill, and can help reduce dependency on foreign fossil fuels and geo-extraction. Pilot Jeremy Roswell plans to make the first flight from Sydney to London using diesel fuel from recycled plastic waste manufactured by Cynar PLC.

Waste Tire Disposal

In the United States alone, over 290 million car tires are discarded annually. Pyrolysis of scrap or waste tires (WT) is an attractive alternative to disposal in landfills, allowing the high energy content of the tire to be recovered as fuel. Using tires as fuel produce equal energy as burning oil and 25% more energy than burning coal.

An average car tire is made up of 50-60% hydrocarbons, resulting in a yield of 38-56% oil, 10-30% gas and 14-56% char. The oil produced is largely composed of benzene, diesel, kerosene, fuel oil and heavy fuel oil, while the produced gas has a similar composition to natural gas. The proportion and the purity of the products are governed by two major factors:

1. Environment (e.g. pressure, temperature, time, reactor type)

2. Material (e.g. age, composition, size, type)

As car tires age, they increase in hardness, making it more difficult for pyrolysis to break the molecules into shorter chains. This shifts the yield composition towards diesel oil which is composed of larger molecules. Conversely, an increase in temperature increases the likelihood of breaking the molecule chain and shifts the yield composition towards benzene oil which is composed of smaller molecules. Other products from car tire pyrolysis include steel wires, carbon black and bitumen.

Although the pyrolysis of WT has been widely developed throughout the world, there are legislative, economic, and marketing obstacles to widespread adoption. Oil derived from tire rubber pyrolysis contains high sulfur content, which gives it high potential as a pollutant and should be desulfurized A number of prototype and full-scale pyrolysis plants specialized in carbon black production have successfully established across the world, including the United States, France, Germany and Japan. Because carbon black is used for pigment, rubber strengthening and UV protection, it is a relatively large and growing market. Pyrolysis plants specialized in fuel oil production is not an implausible concept. However, as profits of such ventures come from the added value between the production and distillation of oil, there is little profit without vertical integration in the oil industry. The inconsistency of the feedstock makes it very difficult to control the uniformity of the products and makes oil companies hesitant to purchase oil produced via pyrolysis. Finally, the cost of producing oil through conventional means is generally less expensive than this alternative. To date, there is no known commercially profitable standalone pyrolysis plant that specializes in oil production. However, with funding to upgrade pyrolysis oil to light fuel grade, this may be possible. Nevertheless, pyrolysis is a valuable method for disposing waste tires.

Chemical Analysis

Pyrolysis can be used for the molecular characterisation of molecules when used in conjunction with gas chromatography-mass spectrometry (Py-GC-MS). This technique has been used to analyse the method and products of fungal decay of wood.

Thermal Cleaning

Pyrolysis is also used for *thermal cleaning*, an industrial application to remove organic substances such as polymers, plastics and coatings from parts, products or production components like extruder screws, spinnerets and static mixers. During the thermal cleaning process, at temperatures between 600 °F to 1000 °F (310 C° to 540 C°), organic material is converted by pyrolysis and oxidation into volatile organic compounds, hydrocarbons and carbonized gas. Inorganic elements remain.

Several types of thermal cleaning systems use pyrolysis:

- *Molten Salt Baths* belong to the oldest thermal cleaning systems; cleaning with a molten salt bath is very fast but implies the risk of dangerous splatters, or other potential hazards connected with the use of salt baths, like explosions or highly toxic hydrogen cyanide gas;

- *Fluidized Bed Systems* use sand or aluminium oxide as heating medium; these systems also clean very fast but the medium does not melt or boil, nor emit any vapors or odors; the cleaning process takes one to two hours;

- *Vacuum Ovens* use pyrolysis in a vacuum avoiding uncontrolled combustion inside the cleaning chamber; the cleaning process takes 8 to 30 hours;

- *Burn-Off Ovens*, also known as *Heat-Cleaning Ovens*, are gas-fired and used in the painting, coatings, electric motors and plastics industries for removing organics from heavy and large metal parts.

Processes

In many industrial applications, the process is done under pressure and at operating temperatures above 430 °C (806 °F). For agricultural waste, for example, typical temperatures are 450 to 550 °C (840 to 1,000 °F).

Processes

Since pyrolysis is endothermic, various methods to provide heat to the reacting biomass particles have been proposed:

- Partial combustion of the biomass products through air injection. This results in poor-quality products.

- Direct heat transfer with a hot gas, the ideal one being product gas that is reheated and recycled. The problem is to provide enough heat with reasonable gas flow-rates.

- Indirect heat transfer with exchange surfaces (wall, tubes): it is difficult to achieve good heat transfer on both sides of the heat exchange surface.

- Direct heat transfer with circulating solids: solids transfer heat between a burner and a pyrolysis reactor. This is an effective but complex technology.

For flash pyrolysis, the biomass must be ground into fine particles and the insulating char layer that forms at the surface of the reacting particles must be continuously removed. The following technologies have been proposed for biomass pyrolysis:

- Fixed beds used for the traditional production of charcoal: poor, slow heat transfer result in very low liquid yields.

- Augers: this technology is adapted from a Lurgi process for coal gasification. Hot sand and biomass particles are fed at one end of a screw. The screw mixes the sand and biomass and conveys them along. It provides a good control of the biomass residence time. It does not dilute the pyrolysis products with a carrier or fluidizing gas. However, sand must be reheated in a separate vessel, and mechanical reliability is a concern. There is no large-scale commercial implementation.

- Electrically heated augers: one process uses an electrical current passed through an auger to heat the material giving excellent heat transfer by contact and radiation to the waste material.

- Ablative processes: biomass particles are moved at high speed against a hot metal surface. Ablation of any char forming at a particle's surface maintains a high rate of heat transfer. This can be achieved by using a metal surface spinning at high speed within a bed of biomass particles, which may present mechanical reliability problems but prevents any dilution of the products. As an alternative, the particles may be suspended in a carrier gas and introduced at high speed through a cyclone whose wall is heated; the products are diluted with the carrier gas. A problem shared with all ablative processes is that scale-up is made difficult, since the ratio of the wall surface to the reactor volume decreases as the reactor size is increased. There is no large-scale commercial implementation.

- Rotating cone: pre-heated hot sand and biomass particles are introduced into a rotating cone. Due to the rotation of the cone, the mixture of sand and biomass is transported across the cone surface by centrifugal force. The process is offered by BTG-BTL, a subsidiary from BTG Biomass Technology Group B.V. in The Netherlands. Like other shallow transported-bed reactors relatively fine particles (several mm) are required to obtain a liquid yield of around 70 wt.%. Larger-scale commercial implementation (up to 5 t/h input) is underway.

- Fluidized beds: biomass particles are introduced into a bed of hot sand fluidized by a gas, which is usually a recirculated product gas. High heat transfer rates from fluidized sand result in rapid heating of biomass particles. There is some ablation by attrition with the sand particles, but it is not as effective as in the ablative processes. Heat is usually provided by heat exchanger tubes through which hot combustion gas flows. There is some dilution of the products, which makes it more difficult to condense and then remove the bio-oil mist from the gas exiting the condensers. This process has been scaled up by companies such as Dynamotive and Agri-Therm. The main challenges are in improving the quality and consistency of the bio-oil.

- Circulating fluidized beds: biomass particles are introduced into a circulating fluidized bed of hot sand. Gas, sand, and biomass particles move together, with the transport gas usually being a recirculated product gas, although it may also be a combustion gas. High heat transfer rates from sand ensure rapid heating of biomass particles and ablation stronger than with regular fluidized beds. A fast separator separates the product gases and vapors from the sand and char particles. The sand particles are reheated in a fluidized burner vessel and recycled to the reactor. Although this process can be easily scaled up, it is rather complex and the products are much diluted, which greatly complicates the recovery of the liquid products.

- Mechanical Fluidized Reactor (MFR). A mechanical stirrer agitates a hot bed of pure char particles into which biomass particles are injected. The stirrer also enhances heat transfer from the reactor wall to the agitated bed. No fluidization gas is required: evolving vapors aerate the bed and greatly reduce the power consumption of the mechanical stirrer. This compact reactor has been used for a mobile pyrolysis plant.

- Chain grate: dry biomass is fed onto a hot (500 °C) heavy cast metal grate or apron which forms a continuous loop. A small amount of air aids in heat transfer and in primary reactions for drying and carbonization. Volatile products are combusted for process and boiler heating.

Use of Vacuum

In vacuum pyrolysis, organic material is heated in a vacuum to decrease its boiling point and avoid adverse chemical reactions. Called flash vacuum pyrolysis, this approach is used in organic synthesis.

Industrial Sources

Many sources of organic matter can be used as feedstock for pyrolysis. Suitable plant material includes greenwaste, sawdust, waste wood, woody weeds; and agricultural sources including nut shells, straw, cotton trash, rice hulls, switch grass; and animal waste including poultry litter, dairy manure, and potentially other manures. Pyrolysis is used as a form of thermal treatment to reduce waste volumes of domestic refuse. Some industrial byproducts are also suitable feedstock including paper sludge and distillers grain.

There is also the possibility of integrating with other processes such as mechanical biological treatment and anaerobic digestion.

Industrial Products

- syngas (flammable mixture of carbon monoxide and hydrogen): can be produced in sufficient quantities to provide both the energy needed for pyrolysis and some excess production

- solid char that can either be burned for energy or be recycled as a fertilizer (biochar).

Fire Protection

Destructive fires in buildings will often burn with limited oxygen supply, resulting in pyrolysis reactions. Thus, pyrolysis reaction mechanisms and the pyrolysis properties of materials are important in fire protection engineering for passive fire protection. Pyrolytic carbon is also important to fire investigators as a tool for discovering origin and cause of fires.

Chemistry

Current research examines the multiple reaction pathways of pyrolysis to understand how to manipulate the formation of pyrolysis' multiple products (oil, gas, char, and miscellaneous chemicals) to enhance the economic value of pyrolysis; identifying catalysts to manipulate pyrolysis reactions is also a goal of some pyrolysis research. Published research suggests that pyrolysis reactions have some dependence upon the structural composition of feedstocks (e.g. lignocellulosic biomass), with contributions from some minerals present in the feedstocks; some minerals present in feedstock are thought to increase the cost of operation of pyrolysis or decrease the value of oil produced from pyrolysis, through corrosive reactions. The low quality of oils produced through pyrolysis can be improved by subjecting the oils to one or many physical and chemical processes, which might drive production costs, but may make sense economically as circumstances change.

Waste Hierarchy

The evaluation of processes that protect the environment alongside resource and energy consumption to most favourable to least favourable actions. The hierarchy establishes preferred program priorities based on sustainability. To be sustainable, waste management cannot be solved only with technical end-of-pipe solutions and an integrated approach is necessary.

The waste management hierarchy indicates an order of preference for action to reduce and manage waste, and is usually presented diagrammatically in the form of a pyramid. The hierarchy captures the progression of a material or product through successive stages of waste management, and represents the latter part of the life-cycle for each product.

The aim of the waste hierarchy is to extract the maximum practical benefits from products and to generate the minimum amount of waste. The proper application of the waste hierarchy can have several benefits. It can help prevent emissions of greenhouse gases, reduces pollutants, save energy, conserves resources, create jobs and stimulate the development of green technologies.

Life-cycle Thinking

All products and services have environmental impacts, from the extraction of raw materials for production to manufacture, distribution, use and disposal. Following the waste hierarchy will generally lead to the most resource-efficient and environmentally sound choice but in some cases refining decisions within the hierarchy or departing from it can lead to better environmental outcomes.

Life cycle thinking and assessment can be used to support decision-making in the area of waste management and to identify the best environmental options. It can help policy makers understand the benefits and trade-offs they have to face when making decisions on waste management strategies. Life-cycle assessment provides an approach to ensure that the best outcome for the environment can be identified and put in place.

It involves looking at all stages of a product's life to find where improvements can be made to reduce environmental impacts and improve the use or reuse of resources. A key goal is to avoid actions that shift negative impacts from one stage to another. Life cycle thinking can be applied to the five stages of the waste management hierarchy.

For example, life-cycle analysis has shown that it is often better for the environment to replace an old washing machine, despite the waste generated, than to continue to use an older machine which is less energy-efficient. This is because a washing machine's greatest environmental impact is during its use phase. Buying an energy-efficient machine and using low- temperature detergent reduce environmental impacts.

The European Union Waste Framework Directive has introduced the concept of life-cycle thinking into waste policies. This duality approach gives a broader view of all environmental aspects and ensures any action has an overall benefit compared to other options. The actions to deal with waste along the hierarchy should be compatible with other environmental initiatives.

European Union Waste Framework Directive

In 1975, The European Union's Waste Framework Directive (1975/442/EEC) introduced for the first time the waste hierarchy concept into European waste policy. It emphasized the importance of waste minimization, and the protection of the environment and human health, as a priority. Following the 1975 Directive, European Union policy and legislation adapted to the principles of the waste hierarchy.

In 1989, it was formalized into a hierarchy of management options in the European Commission's Community Strategy for Waste Management and this waste strategy was further endorsed in the Commission's review in 1996.

In 2008, the European Union parliament introduced a new five-step waste hierarchy to its waste legislation, Directive 2008/98/EC, which member states must introduce into national waste management laws. Article 4 of the directive lays down a five-step hierarchy of waste management options which must be applied by Member States in this priority order.

Waste prevention, as the preferred option, is followed by reuse, recycling, recovery including energy recovery and as a last option, safe disposal.

Challenges for Local and Regional Authorities

The task of implementing the waste hierarchy in waste management practices within a country may be delegated to the different levels of government (national, regional, local) and to other possible actors including industry, private companies and households. Local and regional authorities can be particularly challenged by the following issues when applying the waste hierarchy approach.

- A coherent waste management strategy must be set up

- Separate collection and sorting systems for many different waste streams need to be established.

- Adequate treatment and disposal facilities must be established.

- An effective horizontal co-operation between local authorities and municipalities and a vertical co-operation between the different levels of government, local to regional and when beneficial, also at the national level need to established

- Finding financing for the establishing or upgrading of expensive sustainable waste management infrastructure to address the needs of managing waste

- A lack of data available on waste management strategies must be overcome and monitoring requirements must be met to implement the waste programs

- The enforcement and control of business plans and practices be established and applied to maximize benefits to the environment and human health

- A lack of administrative capacity at the regional and local level. The lack of finances, information, and technical expertise must be overcome for effective implementation and success of the waste management policies.

Source Reduction

Source reduction involves efforts to reduce hazardous waste and other materials by modifying industrial production. Source reduction methods involve changes in manufacturing technology, raw material inputs, and product formulation. At times, the term "pollution prevention" may refer to source reduction.

Another method of source reduction is to increase incentives for recycling. Many communities in the United States are implementing variable-rate pricing for waste disposal (also known as Pay As You Throw - PAYT) which has been effective in reducing the size of the municipal waste stream.

Source reduction is typically measured by efficiencies and cutbacks in waste. Toxics use reduction is a more controversial approach to source reduction that targets and measures reductions in the upstream use of toxic materials. Toxics use reduction emphasizes the more preventive aspects of source reduction but, due to its emphasis on toxic chemical inputs, has been opposed more vigorously by chemical manufacturers. Toxics use reduction programs have been set up by legislation in some states, e.g., Massachusetts, New Jersey, and Oregon. The 3 R's represent the 'Waste Hierarchy' which lists the best ways of managing waste from the most to the least desirable. Many of the things we currently throw away could be reused again with just a little thought and imagination.

Resource Efficiency

Resource efficiency is about maximising the supply of money, materials, staff, and other assets that can be drawn on by a person or organization in order to function effectively, with minimum wasted (natural) resource expenses. It means using the Earth's limited resources in a sustainable manner while minimising environmental impact.

Motivation

A 2014 report by The Carbon Trust suggested that resource challenges are intensifying rapidly - for example, there could be a 40% gap between available water supplies and water needs by 2030, and some critical materials could be in short supply as soon as 2016. These challenges could lead to disruptions to supply, growing regulatory requirements, volatile fluctuation of prices, and may ultimately threaten the viability of existing business models.

Related Concepts

Resource efficiency measures, methods, and aims are quite similar to those of resource productivity/resource intensity and of the slightly more environment-inclined concept of ecological efficiency/eco-efficiency.

Energy Efficiency

Possible Approaches

To achieve and optimize natural resource and energy effiency, several sustainable economical or production schemes have been proposed over the course of the last 50 years: circular economy, cradle-to-cradle- or regenerative design, as well as biomimetics principles, just to name a few. Common to all of them is built-in sustainability, in which (non-renewable) resource-wasting is ruled out by design. They are generally built to be holistic, robustly self-sustaining and respecting the carrying capacity of the economic / ecological system.

Resource Use Measurement and Identification of Hotspots

A key tool in resource efficiency is measuring different aspects of resource use (e.g. carbon footprint, water footprint, land footprint or material use), then identifying 'hot spots' where the most resources are used and/or where there are the best opportunities to reduce this resource use. For example, WRAP has published information on hotspots for 50 grocery products likely to contribute most to the environmental impacts associated with UK household consumption. WRAP have created a range of tools and guides to help improve business resource efficiency.

Initiatives and Programmes

UNEP

UNEP works to promote resource efficiency and *sustainable consumption and production* (SCP) in both developed and developing countries. The focus is on achieving increased understanding and implementation by public and private decision makers, as well as civil society, of policies and actions for resource efficiency and SCP. This includes the promotion of sustainable resource management in a life cycle perspective for goods and services.

Europe 2020

The resource-efficient Europe flagship initiative is part of the Europe 2020 Strategy, the EU's growth strategy for a smart, inclusive and sustainable economy. It supports the shift towards sustainable growth via a resource-efficient, low-carbon economy.

Resource Efficiency in Tomsk Polytechnic University

In October 2012 Tomsk Polytechnic University (TPU) launched the Development Program of Resource Efficient Technologies for the period 2013-2018. That program was presented by TPU in 2009 at the Russian federal competition "National Research University". A key point of the program of TPU was announced the formation of high school as a world-class university-based staffing and development of technologies for resource-efficient economy.

TPU developed educational module "Resource Efficiency", prepared and published a textbook "Principals of resource efficiency", optional subject matter of the same name introduced in the curriculum (for all disciplines and areas of undergraduate).

TPU envisages university development in the field of resource-efficient technologies that unites six research and educational clusters:

1. Safe Environment

    1. Non-destructive testing and diagnostics

    2. Materials for extreme conditions

    3. Domestic and industrial waste recycling

2. Sustainable Energy

    1. High-temperature superconductivity technologies for energy production

    2. Nuclear and hydrogen fuel of the new generation

3. Hybrid simulation in energy production

4. Resource-efficient generation

3. Medical Engineering

1. Bioengineering materials and technologies

2. Radiation technologies in bioengineering

3. Electrophysical biomedical complexes

4. Planet Resources

1. Resource-efficient use of mineral resources

2. Clear water

3. Green chemistry

5. Cognitive Systems and Telecommunications

1. Cognitive software and hardware systems

2. Wireless telecommunication systems and technologies

6. Social Science and Humanities in Engineering

1. Social science and humanities component of engineering

2. Mechanisms of technical innovations initiation and engineering forethought

Resource Efficient Scotland

Resource Efficient Scotland is a Scottish Government funded programme that helps businesses and the public and third sectors save money by using resources more efficiently.

References

- Ratner, Buddy D. (2004). Pyrolytic carbon. In Biomaterials science: an introduction to materials in medicine. Academic Press. p. 171-180. ISBN 0-12-582463-7. Google Book Search. Retrieved 7 July 2011.

- United Nations Environmental Program (2013). "Guidelines for National Waste Management Strategies Moving from Challenges to Opportunities" (PDF). ISBN 978-92-807-3333-4.

- Department for Environment Food and Rural Affairs. "Guidance on Applying the Waste Hierarchy". gov.uk. Retrieved 2016-05-19.

- A Look at Thermal Cleaning Technology". ThermalProcessing.org. Process Examiner. 14 March

2014. Retrieved 4 December 2015.

- Gary Davis & Keith Brown (April 1996). "Cleaning Metal Parts and Tooling" (PDF). Pollution Prevention Regional Information Center. Process Heating. Retrieved 4 December 2015.

- Thomas S. Dwan (2 September 1980). "Process for vacuum pyrolysis removal of polymers from various objects". Espacenet. European Patent Office. Retrieved 26 December 2015.

- "Paint Stripping: Reducing Waste and Hazardous Material". Minnesota Technical Assistance Program. University of Minnesota. July 2008. Retrieved 4 December 2015.

- "Opportunities in a resource constrained world: How business is rising to the challenge". The Carbon Trust. February 2014. Retrieved 22 July 2014.

- Frequently Asked Questions about Biochar | International Biochar Initiative. Biochar-international.org (2013-04-19). Retrieved on 2013-06-01.

- Refining fast pyrolysis of biomass. Thermo-Chemical Conversion of Biomass (Thesis). University of Twente. 2011. Retrieved 2012-05-30.

Policies Related to Waste Management

Every country has specified policies for the management of waste. Some of these policies are solid waste policy in the United States of America, waste management in Turkey and the regional custom of polluter pays principle. The topics discussed in the chapter are of great importance to broaden the existing knowledge on waste management.

Solid Waste Policy

Solid waste policy in India specifies the duties and responsibilities for hygienic waste management for cities and citizens of India. This policy was framed in September 2000, based on the March 1999 Report of the Committee for Solid Waste Management in Class 1 Cities of India to the Supreme Court, which urged statutory bodies to comply with the report's suggestions and recommendations. These also serve as a guide on how to comply with the MSW rules. Both the report and the rules, summarised below, are based on the principle that the best way to keep streets clean is not to dirty them in the first place. So a city without street bins will ultimately become clean and stay clean. They advocate daily doorstep collection of "wet" (food) wastes for composting, which is the best option for India. This is not only because composting is a cost-effective process practiced since old times, but also because India's soils need organic manures to prevent loss of fertility through unbalanced use of chemical fertilizers.

Solid Waste Policy in the United States

Solid waste policy in the United States is aimed at developing and implementing proper mechanisms to effectively manage solid waste. For solid waste policy to be effective, inputs should come from stakeholders, including citizens, businesses, community based-organizations, non governmental organizations, government agencies, universities, and other research organizations. These inputs form the basis of policy frameworks that influence solid waste management decisions. In the United States, the Environmental Protection Agency (EPA) regulates household, industrial, manufacturing and commercial solid and hazardous wastes under the 1976 Resource Conservation and Recovery Act (RCRA). Effective solid waste management is a cooperative effort involving federal, state, regional, and local entities. Thus the RCRA's Solid Waste program section D encourages the environmental departments of each state to develop comprehensive plans to manage nonhazardous industrial and municipal solid waste.

About Solid Waste

Solid waste means any garbage or refuse, sludge from a wastewater treatment plant, water supply treatment plant, or an air pollution control facility and other discarded material, including solid, liquid, semi-solid, or contained gaseous material resulting from industrial, commercial, mining, and agricultural operations, and from community activities. Solid waste does not include solid or dissolved materials in domestic sewage, solid or dissolved materials in irrigation return flows, or industrial discharges. The large scope of the term "solid waste" means that it must be managed in a variety of different ways and that various levels of government employ different policy instruments in order to accomplish this task.

Types

Generally, the term "solid waste" refers to non-hazardous waste. But according to the Resource Conservation and Recovery Act (RCRA) and other state regulations, hazardous waste is also a part of solid waste. Figure 1 provides the solid waste tree diagram. The following section gives a detailed break up on the types of solid wastes.

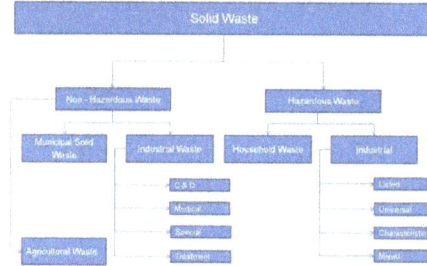

Figure 1 - Solid Waste Tree, Based on Resource Conservation and Recovery Act, United States Environmental Protection Agency

Non-hazardous Solid Wastes

- Municipal Solid Waste: Municipal Solid Waste (MSW), commonly known as trash or garbage, includes all everyday thrown away items from households, commercial and institutional entities, horticulture, and road sweeping. This includes items such as packaging, paper, cardboard, food scraps, plastic bags & containers, glass bottles, grass clippings, furniture, tires, electrical & electronic items, and metals. In 2009, United States residents generated 243 million tons of trash, down from 255 million tons in 2007. In the same period, the per capita generation of MSW lowered to 4.34 lbs/person/day from 4.63 lbs/person/day. The MSW generation trends - total generation and per capita generation - from 1960 to 2009 is provided in Figure 2.

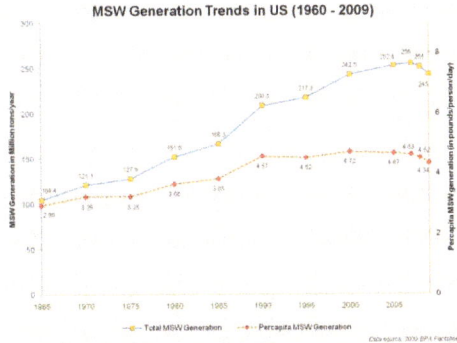

Figure 2 - MSW generation trends (1960–2009), based on US EPA MSW 2009 facts and figures

- Agricultural and Animal Waste: Agricultural wastes include primary crop residues that remain in fields after harvest and secondary processing residues generated from the harvested portions of crops during food, feed, and fiber production. This is generated during the production and distribution through decomposition of food, vegetables, or meat, removal of non-usable parts, removal of substandard products, and spoiling due to substandard packaging. Thus agricultural waste is generated at all stages of food system including farming, storage, processing, and wholesaling. The food scraps generated by retailers and consumers are not included in this category as these scraps enter the waste stream as municipal solid waste that is described in the previous section. Animal wastes are wastes generated from farms and feedlots, also known as Animal Feeding Operations (AFOs) or Concentrated Animal Feeding Operations (CAFOs), consisting of leftover feeds, manure and urine, wastewater, dead animals, and production operation wastes. They produce large amounts of waste in small areas. For example, EPA reports that a single dairy cow produces approximately 120 pounds of wet manure per day equaling to that of 20-40 people. The main problems of animal waste mismanagement are environmental, especially water pollution.

- Industrial Waste: Industrial waste consists of a significant amount of solid waste. EPA reported that each year United States industrial facilities generate and dispose of approximately 7.6 billion tons of industrial solid waste based on 1980s figures. This figure includes waste generated from 17 industrial manufacturers of organic chemicals, inorganic chemicals, iron and steel, plastics and resins, stone, clay, glass, concrete, pulp and paper, food, and kindred products. Industrial waste does not go into the municipal solid waste stream and therefore is landfilled or processed separately. As per EPA guidelines industrial waste management units have to consider waste characterization and minimization methods, waste constituent information factsheets, risk assessment tools, institutional mechanism/stakeholder partnership principles, safe and proper design guidelines, water (surface and ground) and air monitoring procedures, and facility pre- and post-closure recommendations.

- Construction and Demolition Waste: C&D waste includes debris gener-ated during the construction, renovation, and demolition of buildings, roads, and bridges. This can be often bulky and heavy building materials consisting of concrete, building wood waste, asphalt from roads and roof shingles, drywall gypsum, metals, bricks, blocks, glass, plastics, building components like doors, windows, and fixtures, and trees, stumps, earth, and rock from construction and clearing sites. Since this often consists of bulky and heavy materials, proper waste management can improve resources. EPA estimated that 136 million tons of building-related C&D waste was generated in the United States in 1996.

- Treatment Waste: Treatment waste consists of sludge, byproducts, coproducts, or metal scraps resulting from a facility or plant. Sludge is any solid, semisolid, or liquid waste generated from a municipal, commercial, or industrial wastewater treatment plant, water supply treatment plant, or air pollution control facility exclusive of the treated effluent from a wastewater treatment plant. This includes electric arc furnace dust and baghouse dusts. A byproduct is a material that is not a primary product which is not solely or separately produced in a pro-duction process whereas coproducts are intentionally produced. By-products need further processing to be useful whereas coproducts are highly processed and can be sold as a commodity without further pro-cessing. Examples of byproducts include slag, fly ash, heavy ends, dis-tillation column bottoms, etc. and coproducts include metals such as lead produced during the copper refining process. Scrap metal wastes include sheet metal, wire, metal tanks and containers, scrap automo-biles, and machine shop turnings that are generally nonhazardous in nature.

- Medical Waste: Medical waste and biomedical waste consist of all waste materials generated at health care facilities including hospitals, clinics, offices of physicians, dentists, and veterinarians, blood banks, home health care facilities, funeral homes, medical research facilities, and lab-oratories. According to the Medical Waste Tracking Act of 1988, medical waste is:

"Any solid waste that is generated in the diagnosis, treatment, or immuni-zation of human beings or animals, in research pertaining thereto, or in the production or testing of biologicals."

The items include blood-soaked bandages, culture dishes and other glassware, discarded surgical gloves, instruments, lancets, syringes, and needles (medical sharps), cultures, stokes, and swabs used to inoculate cultures, and removed body organs such as tonsils, appendices, etc.

- Special Waste: Six categories of waste were given deferral from hazardous waste requirements by EPA under proposed hazardous waste management regulations. This special category of wastes was maintained until further human health and environmental risk assessments could be completed. As per this deferral, the six categories of special waste are 1) Cement kiln dust, 2) Mining waste, 3) oil and gas drilling muds and oil production brines, 4) beneficiation and processing waste from phosphate rock mining, 5) uranium waste, and 6) utility or fossil fuel combustion waste. The difference between special wastes and other wastes is the large volume of generation of special waste at a time leading to less human and environmental risk.

Hazardous Solid Wastes

The EPA, which regulates hazardous waste under Subtitle C of the Resource Conservation and Recovery Act (RCRA), considers a waste hazardous waste if it is dangerous or potentially harmful to human health or the environment. Hazardous waste can be liquids, solids, gases, or sludges and can be discarded household, industrial, or commercial products such as oil, paints, certain electronics waste, cleaning fluids or pesticides, or the by-products of manufacturing processes.

- Household Hazardous Wastes: This includes used and leftover household products that contain, corrosive, toxic, ignitable, or reactive constituents. Examples are medical waste, used oil, paints, cleaners, batteries, pesticides, and light bulbs/lamps. Since these contain potentially hazardous ingredients, improper disposal can lead to human health risks and environmental pollution. Proper and safe management of hazardous wastes is important in the collection, reuse, recycling, and disposal stages which are mostly facilitated by the municipalities or local governments and specified by EPA in household hazardous waste regulations.

- Industrial Hazardous Wastes: The primary generators of hazardous wastes in any region are industrial facilities, manufacturing and processing units, workshops and maintenance units, nuclear facilities, chemical units, etc. The following section briefly describes the four main types of industrial hazardous wastes.

- Listed Waste: 40CFR Part261 specifies four lists of wastes. These are:

- F-list: This is waste mainly generated from industrial or manufacturing processes or other different industrial sectors, also called non-specific source wastes.

- K-list: This is generated from specific industrial sources such as petroleum refining, wood treatment, pesticide manufacturing, inorganic pigment of chemical manufacturing, metal and coke production, and veterinary pharmaceutical industries.

- P-list and U-list: These are discarded or intended to be discarded commercial chemical products that have listed generic names, container residues, spill residues, or off-specification species. P-list differs from U-waste where the former is acute hazardous waste and the latter toxic waste. This information is available under Hazardous Waste Listings, March 2008.

- Universal Waste: Federal regulations have designated hazardous wastes such as batteries, pesticides, mercury-containing equipment, and light bulbs/lamps as universal wastes. This is a way to streamline them separately and control and facilitate proper collection, storage, recovery or treatment, and disposal that encourages reducing the quantity of such wastes going to landfills and incinerators and thereby increases recovery and recycling rates.

- Characteristic Waste: These are wastes that are defined based on their specific characteristics of ignitability, corrosivity, reactivity, and toxicity. Federal statute 40CFR§261 regulates these wastes. Ignitable wastes are defined by their combustion capacity under conditions when they consist of waste oils and solvents. Corrosive wastes like battery acids are characterized by their pH value – acids (pH ≤ 2) and bases (pH ≥ 12.5). Reactive wastes include lithium-sulfur batteries and explosives that can cause explosions, toxic fumes, or gases and toxic wastes that are harmful to human health or environment when inhaled or ingested or disposed. Examples of toxic wastes include mercury and lead.

- Mixed Waste: These are wastes that contain both radioactive and hazardous waste components making them complicated to regulate. Low Level Mixed Wastes (LLMW) are generated from sources such as industrial, hospital, and nuclear power plant facilities and also from processes such as medical diagnostic testing and research, pharmaceutical and biotechnology development, pesticide research, and nuclear power plant operations. The other two types are High Level Mixed Waste (HLW) and Mixed Trans Uranic Waste (MTRU).

Sources

In 2009, U.S. residents, businesses, and institutions produced more than 243 million tons of municipal solid waste, which is approximately 4.34 lbs/person/day. In addition, American industrial facilities generate and dispose of approximately 7.6 billion tons of industrial solid waste each year as per the EPA estimations in 1980. These levels may be much higher now in the 21st century. The comprehensive report is available at the link: 2009 – MSW facts and figures by the EPA. The primary sources of solid waste are residential, commercial, and industrial entities, construction and renovation sites, hospitals, agricultural fields and animal farms, and treatment and processing plants.

Disposal

Before the 1980s most of the waste generated was either landfilled or burned. More than 90% of the municipal solid waste was landfilled or disposed with less than 7% materials recovery during the 1960s and 70s. This trend started changing after the 1980s when landfill disposal declined to about 54% and resource recovery increased to more than 33%. The most recent numbers from the EPA indicate that in 2012 the US municipal solid waste recycle rate was 34.5%. This section describes the common methods of solid waste disposal practiced in United States and worldwide.

- Landfills: These are technically designed areas where waste is disposed scientifically. They are characterized by liners that prevent seepage of leachates into the groundwater. There are different designs for landfills used for municipal solid waste or household waste, construction & demolition waste, and hazardous waste. According to an EPA report, the number of municipal solid waste landfills has gone down from 7924 in 1988 to 1754 in 2006. There were close to 1900 construction & demolition landfills in 1994.

- Ocean Dumping:

- Combustion or Incineration: Combustion or incineration of waste reduces the amount of landfill space needed by burning waste in a controlled manner and also generates electricity through waste-to-energy technologies such as gasification, pyrolysis, anaerobic digestion, fermentation, etc.

- Transfer Stations: Transfer stations are intermediate facilities where the collected municipal solid waste is unloaded from collection trucks, compacted to reduce the volume of the waste, and held for a short time before it is reloaded onto larger, long-distance trucks or containers for shipment to landfills or other treatment and disposal facilities.

- Recovery & Recycling: Wastes are also good sources of raw materials. Recovery and recycling of wastes can help to reduce the use of virgin materials for producing new goods. Recycling construction & demolition waste can also save the space in landfills and large amounts of materials like metals, glass, plastics, and cardboards can be recovered.

- Composting: Composting is a way to return nutrients back into the environment by allowing microorganisms to turn the waste into manure. Applying this manure to agricultural land can improve the fertility of the soil providing it essential nutrients. It is estimated that 27.8% of the municipal solid waste generated in United States in 2009 was organic waste consisting of yard trimmings and food waste that is compostable. The agricultural and animal waste generated can also be composted and used as manure. For example, it is estimated that a dairy cow produces approximately 40 pounds of waste (dung, urine) per day which can be dried and used as manure. This manure can also be used in digesters to produce biogas (methane gas) or electricity, and other biofuels like ethanol.

Costs and Problems Associated with Waste

Some of the main issues associated with waste are open dumping, odor, particulate matter emissions, leachate seepage from landfills, greenhouse gas (GHG) emissions that lead to air pollution, surface and groundwater pollution, food chain contamination, land area depletion, human health impacts, environmental degradation, and negative impacts on plant and animal life.

Rationales for Solid Waste Policy

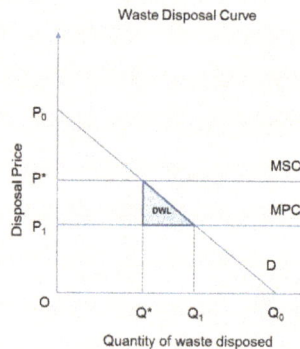

Figure 3 - Demand Curve and Deadweight loss (DWL), Based on Portney and Stavins (2000), Page 269

All levels of government - federal, state, and local - are involved in regulating solid waste in United States. Proper waste management extends from solid waste collection, segregation, transportation, storing, treatment and disposal to education, labeling, trading, and interstate & intercontinental movement of waste. Portney and Stavins (2000) provide the following three rationales for government intervention in private waste markets:

1. Economies of scale - The cost of producing goods or services decreases as production increases. With regards to solid waste this principle applies to landfills where the average cost of landfill construction, operation, and maintenance decreases as waste disposed of increases. This propels interstate trade where private parties divert wastes to large regional landfills.

2. Public bad - Waste creates dissatisfaction to people which reduces social benefits or increases social cost, making it the opposite of a public good. The government, through its policies, makes waste an "excludable" good (or bad) thereby creating opportunities for waste collection firms to charge the household, industrial, and commercial waste generators for proper collection and disposal.

3. Negative externalities - Production of waste leads to environmental pollution especially when it is illegally disposed of, openly dumped, or burned, resulting in groundwater contamination or air pollution. It has been proven that emissions include high amounts of methane and trace amounts of benzene, hydrogen sulphide, and chorinated hydrocarbons along with other gases.

This demand curve and the deadweight loss (DWL) associated with waste disposal (landfilling) is illustrated in Figure 3.

Mechanisms and Policy Framework

The government has a wide variety of different policy tools at its disposal from which it can choose. Due to the diverse nature of solid waste, the government employs a number of different policy tools at various levels in order to ensure efficient and safe handling and disposal of the many different types of waste, as well as in order to encourage recycling and source reduction. The following is a sampling of tools the United States government employs with regards to solid waste.

Deposit Refund

Deposit-refund systems or container deposit legislation, also known as "bottle bills", can be viewed either as a tax on producing waste in the form of beverage containers or as a subsidy for properly recycling these containers. When a retailer buys products from a distributor, it must pay a deposit for each beverage container it purchases. The retailer then includes the cost of the deposit in the item's price, passing it to the consumer. However, the consumer is refunded this money by properly disposing of the used beverage container at a retail or redemption center. The retailer also recoups the deposit from the distributor. This system encourages consumers to properly dispose of the waste they generate by buying beverages in disposable containers. It also creates a privately funded system for the handling of this waste. A deposit-refund bill named National Beverage Container Reuse and Recycling Act was introduced by the House of Representatives in 1994 but never became federal law. Bottle bills are currently in place in ten states as well as in Guam. Delaware repealed its bottle bill in 2010. Oregon was the first state to institute a bottle bill in 1971. The most common deposit is five cents, but this varies by state and by the type of container.

Pay as You Throw

Pay as you throw is a model for pricing the disposal of municipal solid waste by unit of waste rather than by charging a uniform price for pickup and disposal. This acts as a tax on waste - the more waste a household produces, the more it will be charged for its disposal. Pay as you throw is administered on the municipal level. The purpose behind this system is to discourage waste generation and to encourage recycling. By charging citizens per unit of waste, municipalities hope to discourage waste generation by causing households to consider the quantity of waste they are producing by making them pay for it. It is estimated that pay as you throw programs have decreased municipal solid waste by about 17% in weight, with a 6% decrease attributed to source reduction efforts and an 8-11% due to waste diversion to recycling and yard programs. In 2006, pay as you throw had been instituted in over 7,000 United States communities.

Permits

Under the RCRA, the EPA issues permits to ensure the safe treatment, storage, and disposal of hazardous wastes. In order to receive a permit, the party managing the waste has to meet certain criteria, as specified by the Act. Permits are used to set a minimum baseline of safety standards that must be met in the handling and disposal of waste in order to control this process and ensure a degree of safety is achieved. This is an example of command and control regulation, by which the government specifies certain standards that parties must meet.

Technology Standards

Technology standards are another form of command and control regulation by the government. Technology standards stipulate certain types or levels of technologies that must be employed to ensure the safe storage or treatment of waste. For example, technology standards have been created for the design of landfills and there are requirements for the design of the liners in order to prevent leechate.

Performance Standards

Performance standards dictate maximum levels of emissions that may be released in the process of waste management and disposal. These standards are set by the federal government, but can be made more stringent by states. For example, incinerators may not emit over 180 mg of particulate matter per dry standard cubic meter. Other emissions from incinerators are also regulated.

Labeling

Different labeling standards are required by the federal government and by some states for different types of waste such as hazardous waste and medical waste. Labeling ensures that those coming into contact with these types of waste are aware of the nature of the waste. In this way, labeling is also intended to help ensure proper handling and disposal.

Challenges and Goals

The EPA has set forth challenges and goals with regards to solid waste. The Resource Conservation Challenge aims to:

- "Prevent pollution and promote reuse and recycling;

- Reduce priority and toxic chemicals in products and waste; and

- Conserve energy and materials."

It has also issued a challenge to increase recycling to encompass 35% of the country's municipal solid waste. The EPA's other three focuses are on recycling electronics, recycling industrial materials, and reducing priority and toxic chemicals. These challenges and goals are supported by voluntary programs and partnerships.

Partnerships

The EPA has established a number of partnerships with businesses and organizations, industries, states, local governments, tribes, and other entities to reduce and effectively manage waste. Examples of these partnerships are Plug-In To eCycling, the Schools Chemical Cleanout Campaign, and WasteWise. All of these examples aim to meet the goals of the Resource Conservation Challenge. These partnerships are voluntary. Entities may enter into these partnerships because of a variety of expected benefits, including costs savings and improved public image. In another example, EPA and state and tribal representatives jointly developed a framework for industrial waste management aimed to establish a common set of guidelines. Under the Federal Advisory Committee Act, EPA convened a focus group consisting of industry and public stakeholders to provide assistance throughout the industrial waste management process.

Information

On its wastes website, the EPA provides a large amount of information on topics pertaining to waste, such as source reduction and recycling. In this way, the government is working to educate its citizens in order to reduce the amount of waste and ensure its proper disposal in a non coercive manner. This website is also a good source for people looking for instructions on how to properly dispose of items such as compact fluorescent bulbs or electronics.

United States Legislation

History

Federal solid waste law has gone through four major phases. The Solid Waste Disposal Act (SWDA) of 1965 was the first U.S. federal solid waste management law enacted. It focused on research, demonstrations, and training. In a second phase, the Resource

Recovery Act of 1970 emphasized reclaiming energy and materials from solid waste instead of dumping. In a third phase, the federal government started playing more active regulatory role, with the Resource Conservation and Recovery Act (RCRA) of 1976. RCRA instituted the first federal permitting program for hazardous waste and it also made open dumping illegal. RCRA focuses only on active and future facilities and does not address abandoned or historical sites which are managed under the Comprehensive Environmental Response, Compensation, and Liability Act (CERCLA) of 1980 - commonly known as Superfund. Implementation of RCRA was relatively slow and Congress reauthorized and strengthened RCRA through the Hazardous and Solid Waste Amendments (HSWA) of 1984. This was the beginning of the fourth phase. The 1984 RCRA Amendments suggested a policy shift away from land disposal and toward more preventive solutions. RCRA has been amended on two occasions since HSWA: the Federal Facility Compliance Act of 1992 which strengthened enforcement of RCRA at federal facilities and the Land Disposal Program Flexibility Act of 1996 which provided regulatory flexibility for land disposal of certain wastes.

Solid waste legislation has been constantly strengthened and improved by the introduction of amendments to the major laws mentioned above and other specific laws. The most important amendments are:

- Resource Recovery Act of 1970 which provides state and local governments with technical and financial help in planning and developing resource recovery and waste disposal systems

- Used Oil Recycling Act of 1980 which defines the terms used oil, recycled oil, lubricating oil, and re-refined oil, and encourages state to use recycled oil

- Solid Waste Disposal Act Amendments of 1980 which targets against hazardous waste dumping

- Superfund Amendments and Reauthorization Act (SARA) of 1986 which amends CERCLA of 1980, increases state involvement in the Superfund program, and encourages greater citizen participation in decision making

- Medical Waste Tracking Act of 1988 which defines medical waste and introduces management standards for its segregation, packaging, labeling, and storage

- Ocean Dumping Ban Act of 1988 which prohibits all municipal sewage sludge and industrial waste dumping into the ocean

- RCRA cleanup reforms I&II of 1999 and 2001 which accelerate the cleanup of hazardous waste facilities regulated under RCRA

- Used Oil Management Standards of 2003 which defines used oil management standards in a more exact manner

Statutes and rules designed to improve community access to information about chemical hazards:

- Emergency Planning and Community Right-To-Know Act (EPCRA), also known as SARA Title III of 1980, which provides for notification of emergency releases of chemicals, and addresses communities' right-to-know about toxic and hazardous chemicals

- RCRA Expanded Public Participation Rule of 1996 which encourages communities' involvement in the process of permitting hazardous waste facilities and expands public access to information about such facilities

Statutes and rules designed to prevent pollution:

- Pollution Prevention Act of 1990 which requires the EPA to establish an Office of Pollution Prevention and the owners and operators of manufacturing facilities to report annually on source reduction and recycling activities

- Hazardous Waste Combustors; Revised Standards; Final Rule - Part 1 of 1998 which provides for a conditional exclusion from RCRA for fuels which are produced from a hazardous waste and promotes the installation of cost effective pollution prevention technologies

Federal legislation

Resource Conservation and Recovery Act (RCRA)

Within RCRA, the EPA has three comprehensive waste management programs:

1. Subtitle C - Hazardous waste: The Subtitle C program establishes a system for controlling hazardous waste from its generation until its ultimate disposal ("cradle-to-grave" approach). This program identifies the criteria to determine hazardous waste and establishes requirements for all of the parties: producers, transporters and disposal facilities.

2. Subtitle D - Solid waste: The Subtitle D program establishes a system for controlling (primarily non-hazardous) solid waste, such as household waste. The program provides the states and local governments with guidance, policy and regulations for the efficient waste management.

3. Subtitle I - Underground storage tanks: The Subtitle I program in RCRA regulates toxic substances and petroleum products stored in underground storage

tanks (UST). The program establishes requirements for design and operation of UST aimed at preventing accidental spills.

Main RCRA accomplishments: Some of the main achievements of RCRA since its implementation are given below:

- Established design and performance standards for landfills and treatment technologies;

- Established "cradle-to-grave" tracking of hazardous waste;

- Caused the closure of a large number of mismanaged facilities; two-thirds of non-compliant land disposal facilities were closed;

- Prevented the disposal of untreated wastes into and onto the land;

- Permitted more than 900 hazardous waste management facilities;

- Assessed over 1,600 facilities;

- Authorized 48 states for the base RCRA program.

Improvement Areas: There are a number of lessons learned post-implementation of RCRA. First of all it appeared that close Congressional oversight could limit flexibility. After 1984, when HSWA were approved, EPA's discretion was influenced by close Congressional oversight. Congress set specific implementation deadlines for the hazardous waste program. The very demanding regulatory development schedule did not allow EPA to pay enough attention to other very important priorities. The important priorities for EPA found by their study are listed below:

- Program evaluation and long-term priorities should be strengthened;

- Old regulations should be revised; New regulations could make it difficult for state programs;

- Stronger focus on environmental data is needed;

- State authorization should be faster;

- Alternates should be examined;

- Potential regulatory overlaps or inconsistencies should be addressed.

State Legislation

Federal guidelines have provided state, tribal, and some local governments regulatory responsibility for ensuring proper management of wastes generated from each source in their region but the programs might vary considerably in their guidelines and implementation. As of 2011, the EPA has authorized forty-eight states, except Alaska and Iowa (Hawaii was added in 2001), to implement the RCRA, meaning the states' regulations must meet at least the requirements set at the federal level and may be more stringent. Many states follow the federal rules for hazardous waste management and also have more stringent state requirements on hazardous and toxic wastes in particular. California, New York and Iowa are some states that have additional requirements. For example, the California Department of Toxic and Hazardous Substances distinguishes discarded mercury-containing products and waste oil as separate groups of hazardous waste.

Municipal Legislation

Municipalities are in charge of local recycling and trash collection. They can choose whether to contract these services out to private companies or not and how to charge for these services. Municipalities also may adopt approaches of converting waste to energy through methods such as generating electricity from landfill gas. Therefore, municipalities play an important role in everyday waste management. There is, of course, a wide variety of implementation across the country.

Challenges and Issues

Solid waste management challenges and issues that should be considered while framing solid waste policy include proper waste generation, segregation, collection, transportation, and disposal methods, landfill management, hazardous and other toxic material management, treatment, incineration, recycling and other technology standards, monitoring, evaluation, and continuous improvement methods. In addition to these issues, policy has to address the short term and long-term economic, environmental, and social costs and benefits, funding methods, and roles of various stakeholders.

Waste Management in Turkey

Turkey generates 28,858,880 tons of solid municipal waste per year; the annual amount of waste generated per capita amounts to 390 kilograms. According to Waste Atlas, Turkey's waste collection coverage rate is 77%, whereas its unsound waste disposal rate is 69%. While the country has a strong legal framework in terms of laying down common provisions for waste management, the implementation process has been considered slow since the beginning of 1990s.

Overview

Turkey's waste management system is not a priority policy area. The country regardless employs several waste management practices including sanitary landfills, incineration (only for hazardous waste), sterilization, composting, and other advanced disposal methods such as pyrolysis, gasification as well as plasma. The most common method of waste disposal in the country, especially for municipal waste, is landfilling. The municipal waste is collected on a regularly scheduled basis. The metropolitan municipality and other municipalities are responsible for providing collection, transportation, separation, recycling, disposal and storage of waste services.

The management of municipal waste is under the responsibility of municipalities as a regional management approach

The municipal waste is collected on a regularly scheduled basis

Turkey uses a diffuse approach to manage waste by distributing duties and powers among many institutions and organizations.

Legal Framework

Waste management in Turkey is subject to numerous environmental laws. The country had only three laws concerning waste between 1983-2003, whereas ten more regulations were introduced between 2003-2008. Most environmental regulations in Turkey are based on Article 56 of the Constitution, which states:

It is the duty of the State and citizens to improve the natural environment, to protect the environmental health and to prevent environmental pollution.

— Article 56

Turkish Law on Environment no.2872 creates the basis of the legal framework for waste management practices in Turkey:

It is prohibited to discharge all sorts of waste and residue directly or indirectly into receiving environment, storing them or being engaged in a similar activity.

— Article 8

In addition, Law on Amendments in Law on Environment no.5491 (Article 11); Law on Metropolitan Municipalities no.5216 (Article 7); and Municipal Law no.5393 (Article 14 and 15) explain the duties of municipal authorities, whereas Law on Municipal Revenues no. 2464 (Article 97) establishes the polluter-pays principle. Finally, Articles 181 and 182 of the Turkish Penal Code no.5237 (under the section "Crimes Against the Environment") state that intentional pollution of the environment is punishable by law up to five years in prison. The degree of the punishment is decided upon the severity of the pollution and impact on the environment.

Government Efforts

According to the Turkish Ministry of Environment and Urbanization, the management of municipal waste is under the responsibility of municipalities as a regional management approach by the Ministry of Environment and Urbanization. Since 2003, municipalities are implementing municipal waste management projects by cooperating with other municipalities in the region (through the municipalities union). Turkish government drew up a master plan for 2007- 2009 based on the recognition that uncontrolled and unsafe waste disposal is an integral part of daily life in Turkey and poses a serious risk to the environment and to the health of the country's 70 million inhabitants. The number of controlled landfill sites was raised to roughly 3000 - a steep increase on the 90 that existed in the 1990s. As of 2011, there is approximately one sanitary landfill site per municipality.

Ongoing initiatives towards improving the municipal solid waste management in Turkey aimed to set up a waste management system acting in accordance with the related national legislation and EU legislation, covering the establishment of necessary waste treatment facilities (pretreatment facilities and landfills) and transfer stations, reduction of the amount of waste,ensuring recycling and reuse, and reducing the waste transportation costs.

Uncovered landfills remain to be potential sources of flammable biogases, carcirogen and toxic waste, as well as microbial diseases

Lack of public and sectoral environmental consciousness remains to be a problem in Turkey

Waste Mismanagement Practices in Turkey

The main question in the field of waste management is not the legal arrangement itself; but the deficiencies in implementing them. While Turkey uses a diffuse approach to manage its waste, the effectiveness of application process has been negatively affected due to repetitions and gaps in sharing roles and responsibilities among different agencies. This situation, coupled with insufficient institutional capacity and weak technical infrastructure, limits the ability of related legislation to direct the implementation. Turkey is also yet to develop a comprehensive and specific national strategic plan on waste management, or on the climate change overall.

Turkstat reported that waste was Turkey's largest source of methane emissions with 58% between 1990-2011. Eurostat data indicates that Turkey did not recycle any of its municipal solid waste between 2001-2010, although poor reporting, not performance, was given as the cause for the absence of data. The Turkish Ministry of Environment and Urbanization reports the total amount of recycled packaging waste in 2009 to be 2.5 million tonnes, and certainly part of this recycled packaging waste is from MSW sources, but the share is unknown. Out of the approximate 30 million tonnes of municipal waste generated in 2010, 25 million tonnes or 84% were collected and about 98% of this collected waste was landfilled either in sanitary landfills (54%) or dumpsites (44%).

As of 2013, Turkey imposes no landfill tax. According to The Turkish Ministry of Environment and Urbanization, EU Landfill Directive (99/31/EC) will be carried out by no sooner than 2025. In addition, the ministry gave no specific GHG reduction targets in its "Climate Change Action Plan 2011-2023" published in 2012. In the absence of binding international agreements, Turkey is still far from making sound commitments to combat climate change on both national and international level.

Impact of Poor Waste Management

The biggest problem in terms of waste management in the country stems from uncovered landfills, where the garbage is simply left to rot. In 2009, Turkey's overall greenhouse gas emissions increased about 98% compared to 1990 levels. Turkstat data shows that the country's methane emissions increased by 52% between 1990-2011. According to this data, GHG emissions related to waste alone increased by 120% during the mentioned time period. This downward trend, coupled with the country's positive economic and demographic growth since 1990s, creates numerous problems.

Turkey's waste financing system does not take into take into consider the polluter-pays principle sufficiently, so economical tools are weak to deter pollution and financial sources are inadequate for investments. Usage of natural areas (forests, seasides etc.) still causes a great threat to the environment. In addition, insufficient capacity for treatment and disposal of hazardous waste leads to illegal dumping to the nature. Furthermore, recycling rates are poor due to the lack of adequate facilities and incentives in the waste sector. Uncovered landfills remain to be potential sources of flammable biogases, carcirogen and toxic waste, as well as microbial diseases due to the inadequate changes on their status since 1990s. Along with poor funding and reporting, recycling sector in Turkey also suffers from poor environmental consciousness on both public and industrial level.

Polluter Pays Principle

In environmental law, the polluter pays principle is enacted to make the party responsible for producing pollution responsible for paying for the damage done to the natural environment. It is regarded as a regional custom because of the strong support it has received in most Organisation for Economic Co-operation and Development (OECD) and European Community (EC) countries. It is a fundamental principle in US environmental law.

Applications in Environmental Law

The polluter pays principle underpins environmental policy such as an ecotax, which, if enacted by government, deters and essentially reduces greenhouse gas emissions.

Australia

The state of New South Wales in Australia has included the polluter pay principle with the other principles of ecologically sustainable development in the objectives of the Environment Protection Authority.

France

In France, the Charter for the Environment contains a formulation of the polluter pays principle (article 4):

Everyone shall be required, in the conditions provided for by law, to contribute to the making good of any damage he or she may have caused to the environment.

Sweden

The polluter pays principle is also known as extended producer responsibility (EPR). This is a concept that was probably first described by Thomas Lindhqvist for the Swedish government in 1990. EPR seeks to shift the responsibility of dealing with waste from governments (and thus, taxpayers and society at large) to the entities producing it. In effect, it internalised the cost of waste disposal into the cost of the product, theoretically meaning that the producers will improve the waste profile of their products, thus decreasing waste and increasing possibilities for reuse and recycling.

The Organisation for Economic Co-operation and Development defines extended producer responsibility as:

a concept where manufacturers and importers of products should bear a significant degree of responsibility for the environmental impacts of their products throughout the product life-cycle, including upstream impacts inherent in the selection of materials for the products, impacts from manufacturers' production process itself, and downstream impacts from the use and disposal of the products. Producers accept their responsibility when designing their products to minimise life-cycle environmental impacts, and when accepting legal, physical or socio-economic responsibility for environmental impacts that cannot be eliminated by design.

Switzerland

The waste management in Switzerland is based on the polluter pays principle. Bin bags (for municipal solid waste) are taxed with pay-per-bag fees in three quarters of the communes (and the recycling rate doubled in twenty years).

United States

The principle is employed in all of the major US pollution control laws: Clean Air Act, Clean Water Act, Resource Conservation and Recovery Act (solid waste and hazardous waste management), and Superfund (cleanup of abandoned waste sites).

Some eco-taxes underpinned by the polluter pays principle include:

- the Gas Guzzler Tax for motor vehicles

- Corporate Average Fuel Economy (CAFE)- a "polluter pays" fine.

- the Superfund law requires polluters to pay for cleanup of hazardous waste sites, when the polluters can be identified.

In International Environmental Law

In international environmental law it is mentioned in Principle 16 of the Rio Declaration on Environment and Development.

Limitations of Polluter Pays Principle

The US Environmental Protection Agency (EPA) has observed that the polluter pays principle has typically not been fully implemented in US laws and programs. For example, drinking water and sewage treatment services are subsidized and there are limited mechanisms in place to fully assess polluters for treatment costs.

References

- Portney, Paul R. & Stavins, Robert N. (January 2000). "Public Policies for Environmental Protection". Resources for future (2 ed.): 261–286. ISBN 1-891853-03-1.

- Köse, Ömer H.; Ayaz, Sait; Köroğlu, Burak. "Waste Management in Turkey: National Regulations and Evaluation of Implementation Results" (PDF). Sayıştay. Sayıştay. Retrieved 10 March 2015.

- Bakas, Ioannis; Leonidas, Milios. "Municipal waste management in Turkey". EEA. European Environment Agency. p. 11. Retrieved 3 April 2015.

- "Çevre Kanunu no.2872" (PDF). Mevzuat Geliştirme ve Yayın Genel Müdürlüğü. 1983. p. 5913. Retrieved 10 March 2015.

- Köse, Ömer. "Turkish Experience In Auditing Waste Management" (PDF). EUROSAI. EUROSAI/T.C. Sayıştay. Retrieved 5 April 2015.

- "Türkiye Cumhuriyeti İklim Değişikliği Eylem Planı 2011 - 2023" (PDF). T.C. Çevre ve Şehircilik Bakanlığı. T.C. Çevre ve Şehircilik Bakanlığı. Retrieved 5 April 2015.

- Baykan, Barış Gençer. "Türkiye Sera Gazı Salımı Azaltma Taahhüdü Vermekten Kaçınıyor - Araştırma Notu 11/121". BETAM. Bahçeşehir Üniversitesi Ekonomik ve Toplumsal Araştırmalar Merkezi. Retrieved 4 April 2015.

- Gönüllü, Talha (1993). "Çöp Depo Yerlerinde Can Emniyeti ve Halk Sağlığı İle İlgili Tedbirler" (PDF). Çevre Dergisi (9): 9. Retrieved 6 April 2015.

Global Waste Management Movements

Countries with the purpose of recycling and treatment trade in waste. This trade is regarded as global waste trade. Countries trade in waste for treatment and recycling. This trade is regarded global waste trade. The following chapter concentrates on all the movements related to global waste management.

Global Waste Trade

The global waste trade is the international trade of waste between countries for further treatment, disposal, or recycling. Toxic or hazardous wastes are often exported from developed countries to developing countries, also known as countries of the Global South. Therefore, the burden of the toxicity of wastes from Western countries falls predominantly onto developing countries in Africa, Asia, and Latin America. The World Bank Report *What a Waste: A Global Review of Solid Waste Management*, describes the amount of solid waste produced in a given country. Specifically, countries which produce more solid waste are more economically developed and more industrialized. The report explains that "[g]enerally, the higher the economic development and rate of urbanization, the greater the amount of solid waste produced." Therefore, countries in the Global North, which are more economically developed and urbanized, produce more solid waste than Global South countries.

Current international trade flows of waste follow a pattern of waste being produced in the Global North and being exported to and disposed of in the Global South. Multiple factors affect which countries produce waste and at what magnitude, including geographic location, degree of industrialization, and level of integration into the global economy.

Numerous scholars and researchers have linked the sharp increase in waste trading and the negative impacts of waste trading to the prevalence of neoliberal economic policy. With the major economic transition towards neoliberal economic policy in the 1980s, the shift towards "free-market" policy has facilitated the sharp increase in the global waste trade. Henry Giroux, Chair of Cultural Studies at McMaster University, gives his definition of neoliberal economic policy:

"Neoliberalism ...removes economics and markets from the discourse of social obligations and social costs. ...As a policy and political project, neoliberalism is wedded to the privatization of public services, selling off of state functions, deregulation of finance and labor, elimination of the welfare state and unions, liberalization of trade in goods and capital investment, and the marketization and commodification of society."

Given this economic platform of privatization, neoliberalism is based on expanding free-trade agreements and establishing open-borders to international trade markets. Trade liberalization, a neoliberal economic policy in which trade is completely deregulated, leaving no tariffs, quotas, or other restrictions on international trade, is designed to further developing countries' economies and integrate them into the global economy. Critics claim that although free-market trade liberalization was designed to allow any country the opportunity to reach economic success, the consequences of these policies have been devastating for Global South countries, essentially crippling their economies in a servitude to the Global North. Even supporters such as the International Monetary Fund, "progress of integration has been uneven in recent decades"

Specifically, developing countries have been targeted by trade liberalization policies to import waste as a means of economic expansion. The guiding neoliberal economic policy argues that the way to be integrated into the global economy is to participate in trade liberalization and exchange in international trade markets. The claim is that smaller countries, with less infrastructure, less wealth, and less manufacturing ability, should take in hazardous wastes as a way to increase profits and stimulate their economies.

Current Debate Over Global Waste Trade

Arguments Supporting Global Waste Trade

Current supporters of global waste trade argue that importing waste is an economic transaction which can benefit countries with little to offer the global economy. Countries which do not have the production capacity to manufacture high quality products can import waste to stimulate their economy.

Lawrence Summers, former President of Harvard University and Chief Economist of the World Bank, issued a confidential memo arguing for global waste trade in 1991. The memo stated:

"I think the economic logic behind dumping a load of toxic waste in the lowest wage country is impeccable and we should face up to that… I've always thought that countries in Africa are vastly under polluted; their air quality is probably vastly inefficiently low compared to Los Angeles… Just between you and me shouldn't the World Bank be encouraging more migration of the dirty industries to the Least Developed Countries?"

This position, which is mainly motivated by economics and financial profit in particular, demonstrates the main argument for global waste trade. The Cato Institute published an article supporting global waste trade suggesting that "there is little evidence that hazardous wastes, which are often chronic carcinogens, contribute to death rates in developing countries." Elaborating on this point, the article argues that "people in developing countries would rationally accept increased exposure to hazardous pollutants in exchange for opportunities to increase their productivity—and, hence, their income."

Overall, the argument for global waste trade rests largely upon a perception that developing countries need to further their economic development. Supporters suggest that in engaging in global waste trade, developing countries of the Global South will expand their economies and increase profits.

Critiques of Global Waste Trade

Critics of global waste trade claim that lack of regulation and failed policies have allowed developing nations to become toxic dump yards for hazardous waste. The ever-increasing amounts of hazardous waste being shipped to developing countries increases the disproportionate risk that the people in these nations face. Critics of the effects of the global waste trade emphasize the enormous amount of hazardous wastes that people in poorer countries must deal with. They highlight the fact that most of the world's hazardous wastes are produced by Western countries (the United States and Europe), yet the people who suffer negative health effects from these wastes are from poorer countries that did not produce the waste.

Peter Newell, Professor of Development Studies, argues that "environmental inequality reinforces and, at the same time reflects, other forms of hierarchy and exploitation along lines of class, race and gender." Arguing that the detrimental effects of hazardous waste trade affect the disadvantaged more than others, critics of global waste trade suggest that the implications of dumping hazardous waste has significant consequences for People of Color, women, and low-income people in particular.

Critiquing the global waste trade for reproducing inequality on a global scale, many activists, organizers, and environmentalists from regions affected in the Global South have vocalized their disappointment with global waste trade policies. Evo Morales, the first indigenous Amerindian President of Bolivia, argues against the current economic system forcing the exploitation of his country and people. He claims:

"If we want to save the planet earth, to save life and humanity, we have a duty to put an end to the capitalist system. Unless we put an end to the capitalist system, it is impossible to imagine that there will be equality and justice on this planet earth. This is why I believe that it is important to put an end to the exploitation of human beings and to the pillage of natural resources, to put an end to destructive wars for markets and raw materials, to the plundering of energy, particularly fossil fuels, to the excessive consumption of goods and to the accumulation of waste. The capitalist system only allows us to heap up waste.

Jean Francois Kouadio, an African native living near a toxic dump site in the Ivory Coast, explains his experience with the effects of toxic substances lingering throughout his community. With major Western corporations dumping their toxic wastes in the Ivory Coast, Kuoadio has lost two children to the effects of toxic wastes. He describes the loss of his second daughter Ama Grace, and how the doctors "said she suffered from acute glycemia caused by the toxic waste." In addition to critics from the Global South, researchers and scholars in the West have begun critiquing the uneven distribution of negative effects these hazardous waste dumpings are causing. Dorceta Taylor, Professor at the University of Michigan, argues how Women of Color in the United States are disproportionately affected by these policies:

"Women of color have been at the forefront of the struggle to bring attention to the issues that are devastating minority communities - issues such as hazardous waste disposal; exposure to toxins; ...Their communities, some of the most degraded environments ... are repositories of the waste products of capitalist production and excessive consumption. As a result, they have been in the vanguard of the struggle for environmental justice; they are the founders of environmental groups, grassroots activists, researchers, conference organizers, workshop leaders, lobbyists, and campaign and community organizers."

T.V. Reed, Professor of English and American Studies at Washington State University, argues that the correlation between historical colonialism and toxic colonialism is based on perceptions of indigenous land as 'waste.' He argues that Western cultures have deemed indigenous land as "underdeveloped" and "empty," and that the people inhabiting it are therefore less "civilized." Using the historical premises of colonialism, toxic colonialism reproduces these same arguments by defining Global South land as expendable for Western wastes.

Toxic Colonialism

Toxic colonialism, defined as the process by which "underdeveloped states are used as inexpensive alternatives for the export or disposal of hazardous waste pollution by developed states," is the core critique against the global waste trade. Toxic colonialism represents the neocolonial policy which continues to maintain global inequality today through unfair trade systems. Toxic colonialism uses the term colonialism because "the characteristics of colonialism, involving economic dependence, exploitation of labour, and cultural inequality are intimately associated within the new realm of toxic waste colonialism."

Electronic Waste

Electronic waste, also known as e-waste, refers to discarded electrical or electronic devices. A rapidly growing surplus of electronic waste around the world has resulted from quickly evolving technological advances, changes in media (tapes, software, MP3), falling prices, and planned obsolescence. An estimated 50 million tons of e-waste are produced each year, the majority of which comes from the United States and Europe. Most of this electronic waste is shipped to developing countries in Asia and Africa to be processed and recycled.

Various studies have investigated the environmental and health effects of this e-waste upon the people who live and work around electronic waste dumps. Heavy metals, toxins, and chemicals leak from these discarded products into surrounding waterways and groundwater, poisoning the local people. People who work in these dumps, local children searching for items to sell, and people living in the surrounding communities are all exposed to these deadly toxins.

One city suffering from the negative results of the hazardous waste trade is Guiyu, China, which has been called the electronic waste dump of the world. It may be the world's largest e-waste dump, with workers dismantling over 1.5 million pounds of junked computers, cell phones and other electronic devices per year.

Incinerator Ash

Incinerator Ash is the ash produced when incinerators burn waste in order to dispose of it. Incineration has many polluting effects which include the release of various hazardous gases, heavy metals, and sulfur dioxide.

Khian Sea Incident

An example of incinerator ash being dumped onto the Global South from the Global North in an unjust trade exchange is the Khian Sea waste disposal incident. Carrying 14,000 tons of ash from an incinerator in Philadelphia, the cargo ship, Khian Sea, was to dispose of its waste. However, upon being rejected by The Dominican Republic, Panama, Honduras, Bermuda, Guinea Bissau, and the Dutch Antilles, the crew finally dumped a portion of the ash near Haiti. After changing the name of the ship twice to try and conceal the original identity, Senegal, Morocco, Yemen, Sri Lanka, and Singapore still banned the ship's entry. Upon consistent rejections, the ash is believed to have been disposed of in the Atlantic and Indian Oceans. Following this disaster of handling hazardous waste, the Haitian government banned all waste imports leading a movement to recognize all of the disastrous consequences of this global waste trade. Based on the Khian Sea waste disposal incident and similar events, the Basel Convention was written to resist what is known to developing countries as 'toxic colonialism.' It was open for signature in March 1989 and went into effect in May 1992. The U.S. has signed the treaty, but has yet to ratify it.

Chemical Waste

Chemical waste is the excess and unusable waste from hazardous chemicals, mainly produced by large factories. It is extremely difficult and costly to dispose of. It poses many problems and health risks upon exposure, and must be carefully treated in toxic waste processing facilities.

Italy Dumping Hazardous Chemicals in Nigeria

One example of chemical waste being exported from the Global North onto the Global South was the event of an Italian business man seeking to avoid European economic regulations. Allegedly exporting 4,000 tons of toxic waste, containing 150 tons of polychlorinated biphenyls, or PCBs, the Italian businessman made $4.3 million in shipping hazardous waste to Nigeria. The Fordham Environmental Law Review published an article explaining the impacts of the toxic waste imposed on Nigeria in further detail:

"Mislabelling the garbage as fertilizers, the Italian company deceived a retired/illiterate timber worker into agreeing to store the poison in his backyard at the Nigerian river port of Koko for as little as 100 dollars a month. These toxic chemicals were exposed to the hot sun and to children playing nearby. They leaked into the Koko water system resulting in the death of nineteen villagers who ate contaminated rice from a nearby farm."

This is just one example of how the traditional trade flow, from developed Western countries has severely, unfairly, and disproportionately impacted developing countries in the Global South.

Shipbreaking in Asia

Another danger to developing countries is the growing issue of shipbreaking, which is occurring mainly in Asia. Industrialized countries seeking to retire used vessels find it cheaper to send these ships to Asia for dismantling. China and Bangladesh are seen as the two hubs of shipbreaking in Asia. One of the main issues lies in the fact that these ships which are now too aged to continue, were constructed at a time with less environmental regulation. In an environmental fact sheet, researchers demonstrate the immense impact this new toxic trade sector has on workers and the environment. For one, the older ships contain health-damaging substances such as asbestos, lead oxide, zinc chromates, mercury, arsenic, and tributyltin. In addition, shipbreaking workers in China and in other developing countries traditionally lack proper equipment or protective gear when handling these toxic substances.

Impacts of the Global Waste Trade

The global waste trade has had negative effects for many people, particularly in poorer, developing nations. These countries often do not have safe recycling processes or facilities, and people process the toxic waste with their bare hands. Hazardous wastes are often not properly disposed of or treated, leading to poisoning of the surrounding environment and resulting in illness and death in people and animals. Many people have experienced illnesses or death due to the unsafe way these hazardous wastes are handled.

Effects Upon the Environment

The hazardous waste trade has disastrous effects upon the environment and natural ecosystems. Various studies explore how the concentrations of persistent organic pollutants have poisoned the areas surrounding the dump sites, killing numerous birds, fish, and other wildlife. There are heavy metal chemical concentrations in the air, water, soil, and sediment in and around these toxic dump areas, and the concentration levels of heavy metals in these areas are extremely high and toxic.

Implications for Human Health

The hazardous waste trade has serious damaging effects upon the health of humans. People living in developing countries may be more vulnerable to the dangerous effects of the hazardous waste trade, and are particularly at risk from developing health problems. The methods of disposal of these toxic wastes in developing countries expose the general population (including future generations) to the highly toxic chemicals. These toxic wastes are often disposed of in open landfills, burned in incinerators, or in other dangerous processes. Workers wear little to no protective gear when processing these toxic chemicals, and are exposed to these toxins through direct contact, inhalation, contact with soil and dust, as well as oral intake of contaminated locally produced food and drinking water. Health problems resulting from these hazardous wastes affect humans by causing cancers, diabetes, alterations in neurochemical balances, hormone disruptions from endocrine disruptors, skin alterations, neurotoxicity, kidney damage, liver damage, bone disease, emphysema, ovotoxicity, reproductive damage, and many other fatal diseases. The improper disposal of these hazardous wastes creates fatal health problems, and is a serious public health risk.

International Responses to Global Waste Trade Issues

There have been various international responses to the problems associated with the global waste trade and multiple attempts to regulate it for over thirty years. The hazardous waste trade has proven difficult to regulate as there is so much waste being traded, and laws are often difficult to enforce. Furthermore, there are often large loopholes in these international agreements that allow countries and corporations to dump hazardous wastes in dangerous ways. The most notable attempt to regulate the hazardous waste trade has been the Basel Convention.

International Treaties and Relevant Trade Law

Basel Convention

The Basel Convention on the Control of Transboundary Movements of Hazardous Wastes and Their Disposal, usually known as the Basel Convention, is an international treaty that plays a crucial role in regulating the transnational movement of hazardous wastes. The Basel Convention was created in 1989 and attempts to regulate the hazardous waste trade, specifically to prevent the dumping of hazardous waste from more developed countries into less developed countries. The Basel Convention was developed following a series of high-profile cases in which large amounts of toxic waste were dumped into less developed countries, poisoning the people and environment. The Convention seeks to reduce the creation of hazardous wastes, and to control and reduce its trade across borders.

The Convention was opened for signatures on 22 March 1989, and officially entered into force on 5 May 1992. As of May 2014, 180 states and the European Union are parties to the Convention. Haiti and the United States have signed the Convention but not ratified it.

ENFORCE

The Environmental Network for Optimizing Regulatory Compliance on Illegal Traffic (ENFORCE) is an agency staffed by relevant experts to promote compliance with the Basel Convention. It is an international body created to deal with transboundary issues of the international hazardous waste trade. Because the issue of the transnational hazardous waste trade crosses many borders and affects many nations, it has been important to have a multinational, multilateral organization presiding over these affairs. The members of ENFORCE include one representative from each of the five United Nations regions that are parties to the Convention as well as five representatives from the Basel Convention regional and coordinating centers, based on equitable geographical representation. Members of organizations such as the United Nations Environmental Programme (UNEP), International Criminal Police Organization (INTERPOL), NGOs working to prevent and stop illegal traffic such as the Basel Action Network (BAN), and many other organizations are also eligible to become members of ENFORCE.

Protocol on Liability and Compensation

In 1999 the Basel Convention passed the Protocol on Liability and Compensation that sought to improve regulatory measures and better protect people from hazardous waste. The Protocol on Liability and Compensation attempts to "assign appropriate liability procedures when the transboundary movements of hazardous wastes result in damages to human health and the environment". The Protocol "imposes strict liability for damages in situations involving Parties to the Basel Convention, but only while they maintain control of the hazardous waste through their respective notifying, transporting, or disposing entities." It seeks to regulate and ensure countries' and corporations' compliance with the Basel Convention laws. However, this Protocol remains unsigned by most countries, so its applicability is limited.

Lomé IV Convention and Cotonou Agreement

In an effort to protect themselves against unfair hazardous waste dumping, the African, Caribbean, and Pacific States (ACP) signed the Lome IV Convention, which is a supplement to the Basel Convention and prohibits the "export of hazardous wastes from the European Community to ACP States." This Convention is one attempt by developing countries to protect themselves from Western countries exporting their waste to poorer nations through the hazardous waste trade. When the Lomé IV Convention expired in 2000, the ACP countries and the European countries entered into a new agreement known as the Cotonou Agreement, which "recognizes the existence of disproportionate risks in developing countries and desires to protect against inappropriate hazardous waste shipments to these countries."

The Bamako Convention

In 1991 multiple developing nations in Africa met to discuss their dissatisfaction with the Basel Convention in regulating the dumping of hazardous waste into their countries, and designed a ban on the import of hazardous wastes into their countries called the Bamako Convention. The Bamako Convention is different from the Basel Convention in that Bamako "essentially bans the import of all hazardous waste generated outside of the OAU [the Organization of African Unity] for disposal or recycling and deems any import from a non-Party to be an illegal act." However, these countries could not effectively implement the stipulations of the Convention and could not prevent the dump of toxic wastes due to limited resources and a lack of powerful enforcement. Therefore, the application of the Bamako Convention was very limited.

Critiques of These Responses

Laura Pratt, expert on the hazardous waste trade, claims that despite local and international attempts to regulate the hazardous waste trade, the "current international agreements, both the widespread, legally binding agreements and the ad hoc agendas among smaller groups of countries, have not been as successful at eliminating toxic waste colonialism as proponents would have hoped." She explains that there are various loopholes in the current system that allow toxic waste to continue being dumped, and toxic colonialism to go unchecked. Some of the problems with these international agreements include continued illegal shipments and unclear definitions of terms.

Fraudulent Shipments and Concealments

Pratt explains that despite attempts to regulate illegal dumping, "[o]ftentimes hazardous waste is simply moved under false permits, bribes, improper labels, or even the pretext of 'recycling,' which is a growing trend." Companies often export their hazardous wastes to poorer countries through illegal smuggling. Attempts to regulate this market have been hindered by a lack of ability to monitor the trade, as many countries do not have any authoritative legislative bodies in place to prevent or punish the illegal trafficking of hazardous wastes. Furthermore, Pratt explains that without coordinated international methods to enforce the regulations, it is extremely difficult for countries to "control the illegal trade of hazardous waste, due to the disparity between enforcement resources and regulation uniformity." Developing nations still bear the brunt of this illegal activity the most, and often do not have the resources or capability to protect themselves.

Issues with Legal Definitions

Another issue with the Basel Convention and other international agreements to regulate the waste trade is the difficulty of establishing clear, uniform definitions regarding wastes. These overly broad and ambiguous definitions cause problems with the international agreements, as different parties interpret the language of the agreements differently and thus act accordingly. For example, the "'lack of distinction between 'waste' and 'products' in the convention and its vague criteria for 'hazardous' allowed the continued export of 'hazardous waste' under the label of commodities or raw materials, despite the fact that these wastes still present environmental and health risks to developing countries."

Basel Convention

The Basel Convention on the Control of Transboundary Movements of Hazardous Wastes and Their Disposal, usually known as the Basel Convention, is an international treaty that was designed to reduce the movements of hazardous waste between nations, and specifically to prevent transfer of hazardous waste from developed to less developed countries (LDCs). It does not, however, address the movement of radioactive waste. The Convention is also intended to minimize the amount and toxicity of wastes generated, to ensure their environmentally sound management as closely as possible to the source of generation, and to assist LDCs in environmentally sound management of the hazardous and other wastes they generate.

The Convention was opened for signature on 31 April 1989, and entered into force on 5 May 1992. As of July 2016, 183 states and the European Union are parties to the Convention. Haiti and the United States have signed the Convention but not ratified it.

History

With the tightening of environmental laws (for example, RCRA) in developed nations in the 1970s, disposal costs for hazardous waste rose dramatically. At the same time, globalization of shipping made transboundary movement of waste more accessible, and many LDCs were desperate for foreign currency. Consequently, the trade in hazardous waste, particularly to LDCs, grew rapidly.

One of the incidents which led to the creation of the Basel Convention was the *Khian Sea* waste disposal incident, in which a ship carrying incinerator ash from the city of Philadelphia in the United States dumped half of its load on a beach in Haiti before being forced away. It sailed for many months, changing its name several times. Unable to unload the cargo in any port, the crew was believed to have dumped much of it at sea.

Another is the 1988 Koko case in which 5 ships transported 8,000 barrels of hazardous waste from Italy to the small town of Koko in Nigeria in exchange for $100 monthly rent which was paid to a Nigerian for the use of his farmland.

These practices have been deemed "Toxic Colonialism" by many developing countries.

At its most recent meeting, 27 November – 1 December 2006, the Conference of the parties of the Basel Agreement focused on issues of electronic waste and the dismantling of ships.

According to Maureen Walsh, only around 4% of hazardous wastes that come from OECD countries are actually shipped across international borders. These wastes include, among others, chemical waste, radioactive waste, municipal solid waste, asbestos, incinerator ash, and old tires. Of internationally shipped waste that comes from developed countries, more than half is shipped for recovery and the remainder for final disposal.

Increased trade in recyclable materials has led to an increase in a market for used products such as computers. This market is valued in billions of dollars. At issue is the distinction when used computers stop being a "commodity" and become a "waste".

As of July 2016, there are 184 parties to the treaty, which includes 181 UN member states, the Cook Islands, the European Union, and the State of Palestine. The 12 UN member states that are not party to the treaty are Angola, East Timor, Fiji, Grenada, Haiti, San Marino, Sierra Leone, Solomon Islands, South Sudan, Tuvalu, United States, and Vanuatu.

Definition of Hazardous Waste

4.5-volt, D, C, AA, AAA, AAAA, A23, 9-volt, CR2032 and LR44 cells are all recyclable in most countries.

Several sizes of button and coin cell. 2 9v batteries were added as a size comparison. Enlarge to see the button and coin cells' size code markings. They are all recyclable in both the UK and Ireland amongst others.

A waste falls under the scope of the Convention if it is within the category of wastes listed in Annex I of the Convention and it exhibits one of the hazardous characteristics contained in Annex III. In other words, it must both be listed and possess a characteristic such as being explosive, flammable, toxic, or corrosive. The other way that a waste may fall under the scope of the Convention is if it is defined as or considered to be a hazardous waste under the laws of either the exporting country, the importing country, or any of the countries of transit.

The definition of the term disposal is made in Article 2 al 4 and just refers to annex IV, which gives a list of operations which are understood as disposal or recovery. The examples of disposal are broad and include also recovery, recycling.

Alternatively, to fall under the scope of the Convention, it is sufficient for waste to be included in Annex II, which lists other wastes, such as household wastes and residue that comes from incinerating household waste.

Radioactive waste that is covered under other international control systems and wastes from the normal operation of ships are not covered.

Annex IX attempts to define "commodities" which are not considered wastes and which would be excluded.

Obligations

In addition to conditions on the import and export of the above wastes, there are stringent requirements for notice, consent and tracking for movement of wastes across national boundaries. It is of note that the Convention places a general prohibition on the exportation or importation of wastes between Parties and non-Parties. The exception to this rule is where the waste is subject to another treaty that does not take away from the Basel Convention. The United States is a notable non-Party to the Convention and has a number of such agreements for allowing the shipping of hazardous wastes to Basel Party countries.

The OECD Council also has its own control system that governs the trans-boundary movement of hazardous materials between OECD member countries. This allows, among other things, the OECD countries to continue trading in wastes with countries like the United States that have not ratified the Basel Convention.

Parties to the Convention must honor import bans of other Parties.

Article 4 of the Basel Convention calls for an overall reduction of waste generation. By encouraging countries to keep wastes within their boundaries and as close as possible to its source of generation, the internal pressures should provide incentives for waste reduction and pollution prevention. Parties are generally prohibited from exporting covered wastes to, or import covered waste from, non-parties to the convention.

The Convention states that illegal hazardous waste traffic is criminal but contains no enforcement provisions.

According to Article 12, Parties are directed to adopt a protocol that establishes liability rules and procedures that are appropriate for damage that comes from the movement of hazardous waste across borders.

Current consensus is that as space is not classed as a "country" under the specific definition, export of e-waste to non terrestrial locations would not be covered. This has been suggested (somewhat laughably) as a way to deal with the "Fridge Mountain" and related deposits of waste in the UK and elsewhere in the event of a way to cheaply access space such as an orbital tether being built.

Basel Ban Amendment

After the initial adoption of the Convention, some least developed countries and environmental organizations argued that it did not go far enough. Many nations and NGOs argued for a total ban on shipment of all hazardous waste to LDCs. In particular, the original Convention did not prohibit waste exports to any location except Antarctica but merely required a notification and consent system known as "prior informed consent" or PIC. Further, many waste traders sought to exploit the good name of recycling and begin to justify all exports as moving to recycling destinations. Many believed a full ban was needed including exports for recycling. These concerns led to several regional waste trade bans, including the Bamako Convention.

Lobbying at the 1995 Basel conference by LDCs, Greenpeace and several European countries such as Denmark, led to the adoption of an amendment to the convention in 1995 termed the Basel Ban Amendment to the Basel Convention. The amendment has been accepted by 86 countries and the European Union, but has not entered into force (as that requires ratification by 3/4 of the member states to the Convention). The Amendment prohibits the export of hazardous waste from a list of developed (mostly OECD) countries to developing countries. The Basel Ban applies to export for any reason, including recycling. An area of special concern for advocates of the Amendment was the sale of ships for salvage, shipbreaking. The Ban Amendment was strenuously opposed by a number of industry groups as well as nations including Australia and Canada. The number of ratification for the entry-into force of the Ban Amendment is under debate: Amendments to the convention enter into force after ratification of "three-fourths of the Parties who accepted them" [Art. 17.5]; so far, the Parties of the Basel Convention could not yet agree whether this would be three fourth of the Parties that were Party to the Basel Convention when the Ban was adopted, or three fourth of the current Parties of the Convention. The status of the amendment ratifications can be found on the Basel Secretariat's web page. The European Union fully implemented the Basel Ban in its Waste Shipment Regulation (EWSR), making it legally binding in all EU member states.

Norway and Switzerland have similarly fully implemented the Basel Ban in their legislation.

In the light of the blockage concerning the entry into force of the Ban amendment, Switzerland and Indonesia have launched a "Country-led Initiative" (CLI) to discuss in an informal manner a way forward to ensure that the transboundary movements of hazardous wastes, especially to developing countries and countries with economies in transition, do not lead to an unsound management of hazardous wastes. This discussion aims at identifying and finding solutions to the reasons why hazardous wastes are still brought to countries that are not able to treat them in a safe manner. It is hoped that the CLI will contribute to the realization of the objectives of the Ban Amendment. The Basel Convention's website informs about the progress of this initiative

References

- Grossman, Gene M.; Krueger, Alan B. (1994). "Environmental Impacts of a North American Free Trade Agreement". In Garber, Peter. The U.S. Mexico Free Trade Agreement. MIT Press. pp. 13–56. doi:10.3386/w3914. ISBN 0-262-07152-5.

- Cunningham, William P & Mary A (2004). Principles of Environmental Science. McGrw-Hill Further Education. p. Chapter 13, Further Case Studies. ISBN 0072919833.

Permissions

Index

www.ingramcontent.com/pod-product-compliance
Lightning Source LLC
Chambersburg PA
CBHW061931190326
41458CB00009B/2716